Rio Grande Rift

El Paso, Texas to Denver, Colorado
July 20–30, 1989

Field Trip Guidebook T318

Leaders:
Jonathan F. Callender *John W. Hawley* *G. Randy Keller*

Contributions By:
Richard M. Chamberlin *Charles F. Kluth* *Kenneth H. Olsen*
Richard P. Lozinsky *Michael McCurry*

American Geophysical Union, Washington, D.C.

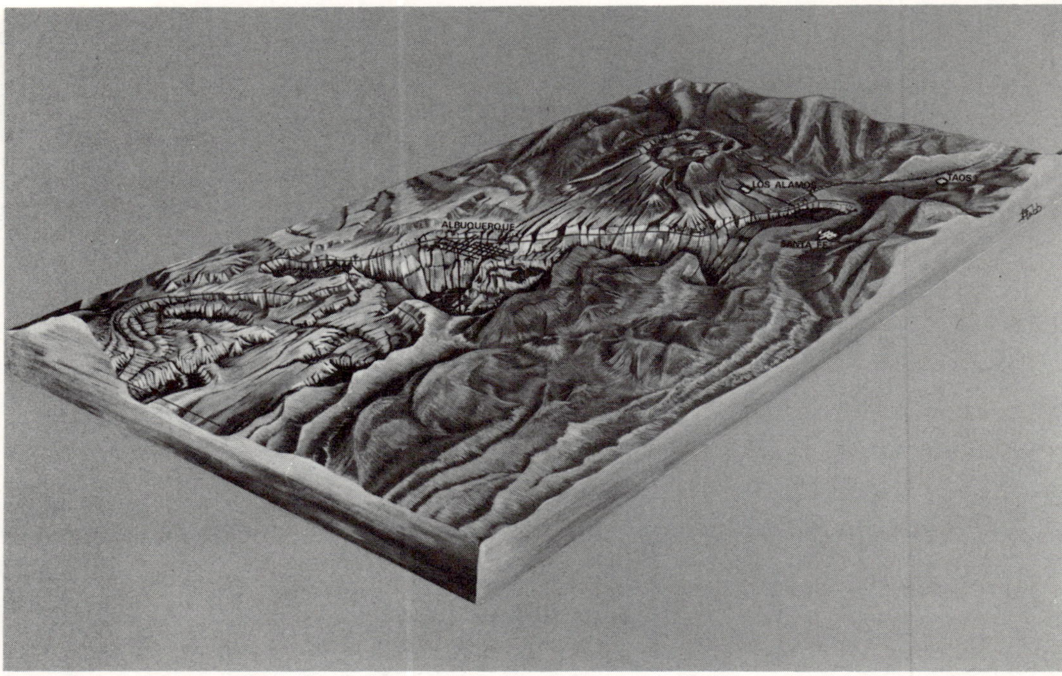

COVER Artist's view of central Rio Grande rift stripped of its sediments and sedimentary rocks.

Leaders:

Jonathan F. Callender
New Mexico Museum of Natural History
P.O. Box 7010
Albuquerque, NM 87194

John W. Hawley
New Mexico Bureau of Mines and Mineral Resources
Socorro, NM 87801

G. Randy Keller
Department of Geological Sciences
University of Texas at El Paso
El Paso, TX 79968

2000 Florida Ave., N.W., Washington, D.C. 20009

ISBN: 0-87590-643-5

Printed in the United States of America

IGC FIELD TRIP T318: RIO GRANDE RIFT

G. Randy Keller,[1] Jonathan F. Callender,[2] John W. Hawley and Richard M. Chamberlin,[3] Charles F. Kluth,[4] Kenneth H. Olsen,[5] and Richard P. Lozinsky[6]

GEOPHYSICAL AND TECTONIC SETTING OF THE RIO GRANDE RIFT

Introduction

It was only in the 1970s that the Rio Grande rift came to be recognized as a coherent tectonic feature which was more than simply an arm of the Basin and Range province (e.g., Chapin, 1971, 1988; Decker and Smithson, 1975; Cordell, 1978). The international scientific community was largely unaware of this feature until an international rift conference was convened in Santa Fe, New Mexico, in the fall of 1978. The volume published as a result of this conference (Riecker, 1979) estab lished the Rio Grande rift as an important Cenozoic Continental rift, and the guidebook produced for this conference (Hawley, 1978) is the basis of this field trip. Although many questions remain, the many studies conducted since 1978 provide an improved picture of the nature and evolution of this rift which extends from northern Colorado into northern Mexico and is superficially similar to the Kenyan arm of the East African rift (Fig. 1). The discussions which follow are intended to augment the papers in Riecker (1979) and briefly review results obtained since its publication. Other recent volumes on the rift include a special section of the Journal of Geophysical Research (May 1986), New Mexico Geological Society Guidebooks for the 1979, 1980, 1982, 1983, 1984, and 1986 field conferences, and field trip guides by Gile et al. (1981), McLemore and Bowie (1987), and Menges (1987).

[1]Department of Geological Sciences, University of Texas at El Paso, El Paso, Texas 79968.

[2]New Mexico Museum of Natural History, P.O. Box 7010, Albuquerque, New Mexico 87194-7010.

[3]New Mexico Bureau of Mines and Mineral Resources.

[4]Chevron Overseas Petroleum, Inc., 6001 Bollinger Canyon Road, San Ramon, California 94583-2324.

[5]Los Alamos National Laboratory, Geophysics, ESS-3, MSC335, Los Alamos, New Mexico 87545.

[6]Leighton and Associates, 667 Brea Canyon Road, Suite 31, Walnut, California 91789.

Evolution

The tectonic evolution of the Rio Grande rift is difficult to study because this feature occurs in a region of complex Cenozoic deformation (Fig. 2). Laramide compression extended up until about 40 Ma at least in the area of the southern rift (e.g., Seager and Mack, 1986). This activity was partially synchronous with a period of widespread area volcanism that extended up until about 30 Ma. Recent studies (Morgan and Golombek, 1984; Seager et al., 1984; Morgan et al., 1986a) have documented two phases of extension beginning at about 30 Ma. The first phase was probably characterized by low-angle normal faulting, NE-SE (±25°) extension, broad relatively shallow basins, and regional extension on the order of 30-50%. This phase extended until about 18 Ma and was followed by a period of relative quiescence. The second phase of extension began about 10 Ma and was characterized by high-angle faulting, E-W extension, grabens and half-grabens which often cut across early features, relatively minor regional extension, and regional uplift of 1 km. Tectonism and volcanism have extended to the present, but the activity seems to be at a low level.

The uplift history which accompanied this extension is very difficult to unravel (Morgan et al., 1986a). Recent uplift of about 1 km is suggested by uplifted erosion surfaces (Scott, 1975; Taylor, 1975), fossil flora (Axelrod and Bailey (1976), and fission track studies (Kelley and Duncan, 1984, 1986). However, this uplift seems to be regional in nature including the Colorado Plateau, southern Rocky Mountains, and Great Plains (Morgan and Swanberg, 1985; Eaton, 1986).

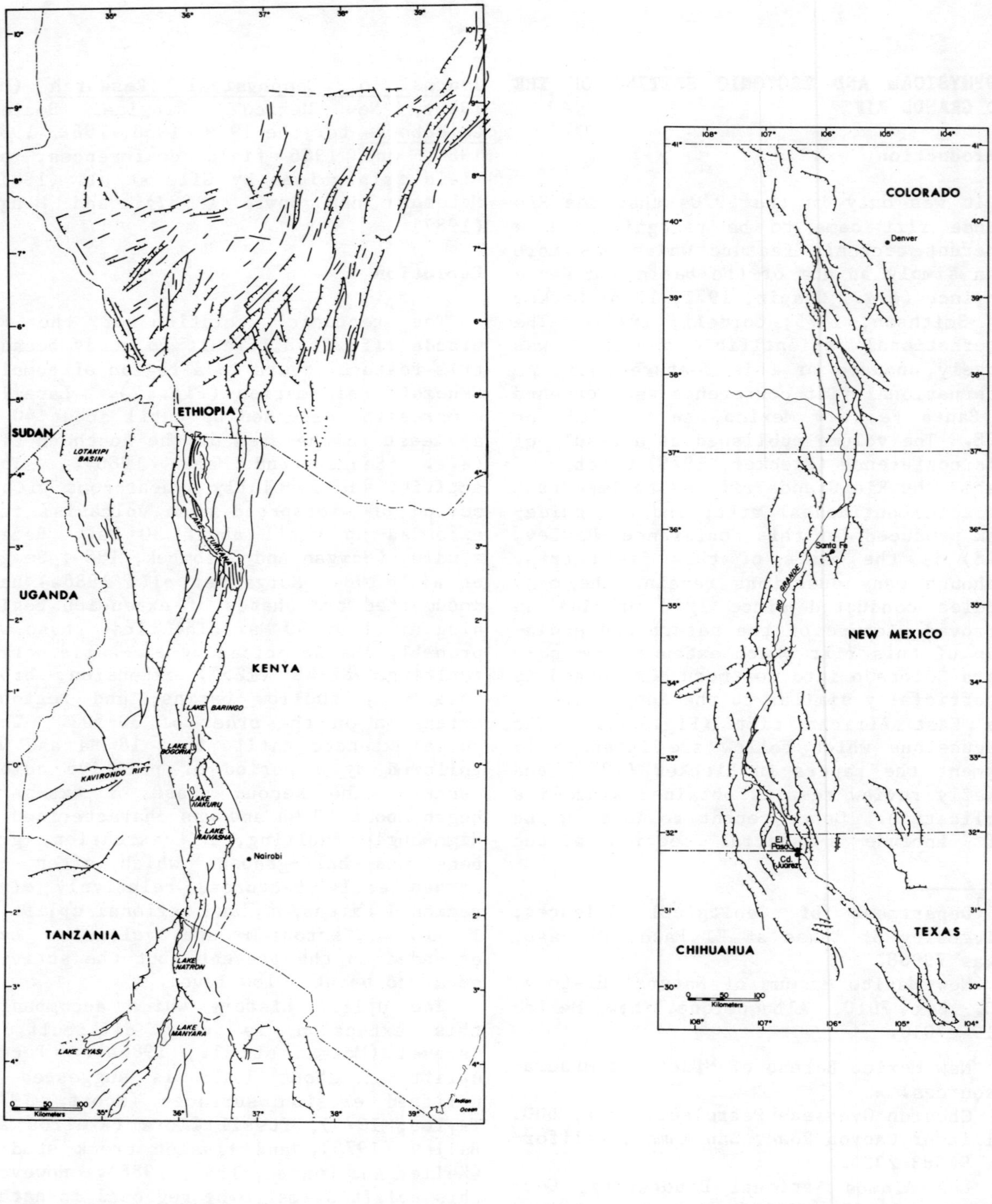

FIGURE 1 A comparison of the generalized patterns of faulting in the Rio Grande rift and Kenyan arm of the East African rift. The scale is the same for each rift.

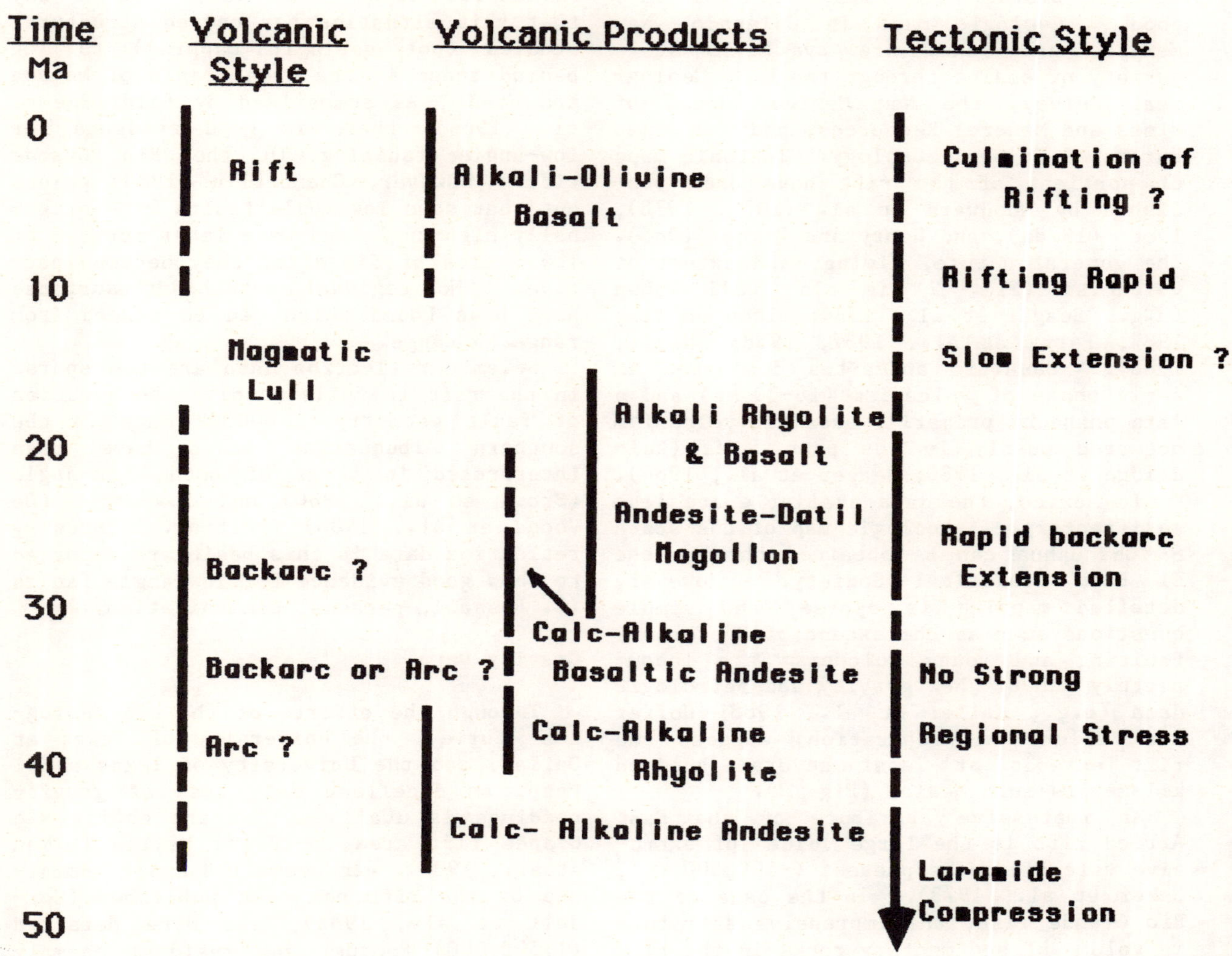

FIGURE 2 Tectonic evolution of the Rio Grande rift area (modified after Paul Morgan, personal communication).

Geologic Data and Structure

From the standpoint of surface mapping, the geologic data set for the portion of the Rio Grande rift in the U.S. is very good. Geologic maps in Colorado, New Mexico, and Texas are available at a variety of scales through the U.S. Geological Survey, the New Mexico Bureau of Mines and Mineral Resources, and the Texas Bureau of Economic Geology. Tectonic maps of portions of the rift have been published by Woodward et al. (1975, 1978), Tweto (1978a), and Henry and Price (1985). The general nature, timing, and extent of volcanism (Fig. 2) is also well known (e.g., Seager et al., 1984; Olsen et al., 1987; Perry et al., 1987, 1988; Chapin, 1988). Basaltic andesites dominated an early phase of volcanism (18-30 Ma) and a late phase of primarily basaltic volcanism occurred mostly in the past 5 Ma (Baldridge et al., 1980; Seager et al., 1984).

In Mexico, the data available are less satisfactory. A geologic map of the state of Chihuahua can be obtained through the El Paso Geological Society. However, detailed mapping is sparse, and simple questions such as the extent of Quaternary faulting and young volcanism are incompletely known. New gravity and petrologic data (e.g., Keller et al., 1988; Hoffer and Keller, in preparation) suggest the rift extends at least several hundred kilometers into Mexico (Fig. 3).

An impressive attribute of the East Africa rift is the large volume of extrusive volcanic rocks present (~500,000 km^3, Baker et al., 1973). In the case of the Rio Grande rift, the impressive attribute is volume of sedimentary rocks in the rift basins. Thanks to the cooperation of the oil industry, we have the most reliable data in the Albuquerque and San Luis basins. In the Albuquerque basin, the Shell Isleta #2 well was drilled to a depth of 6,482 m without encountering Cretaceous strata (Lozinsky, 1988). The Neogene sedimentary rocks alone were over 4.4 km in thickness. In the San Luis basin of southern Colorado, seismic, drilling, and gravity data (Keller et al., 1984; Kluth and Schaftenaar, 1989) indicate that the total thickness of Tertiary sediments is approximately 6.4 km. However, it is difficult to determine what portion is Neogene in age. A variety of geophysical measurements place the thickness of Tertiary sediments in Hueco basin east of El Paso, Texas, at about 3 km (Mattick, 1967; Wen, 1983).

All of these sediments pose a problem from a structural point of view because many key areas are buried. For example, the San Luis basin contains a major buried structural high. It seems fair to say that this situation has caused structural studies (not geologic mapping) to lag behind those in the Great basin of Nevada and Utah. As summarized by Baldridge et al. (1984), there is good evidence for low-angle faulting in the Rio Grande rift. However, Chamberlin (1983) points out that some low-angle faults were originally high-angle and were later rotated to low angles of dip after they became inactive. No regional detachment surfaces have been found which can be traced from range to range.

Seismic reflection data are too sparse in the rift to fully resolve the question of fault geometry. COCORP data in the southern Albuquerque basin have been interpreted in terms of both high-angle (Brown et al., 1980) and low-angle (De Voogd et al., 1986) faulting. Industry reflection data in this basin are reported to show good evidence for low-angle faults (L. Russell, personal communication).

Gravity Data

Through the efforts of the U.S Geological Survey, the University of Texas at Dallas, and the University of Texas at El Paso, an excellent data base of gravity readings is available for the entire Rio Grande rift area including Mexico (Aiken et al., 1988). An overall Bouguer anomaly map of the rift has been published (Cordell et al., 1982), and more detailed (1:500,000) Bouguer and residual anomaly maps of the state of New Mexico have also been published (Keller and Cordell, 1984; Aiken et al., 1978).

The gravity field in the area of the rift is complicated by the presence of large regional elevation variations which are not directly related to the rifting process. These elevation changes and large older crustal features such as early Tertiary batholiths produce regional gravity anomalies which approach 200 milligals. These regional anomalies complicate interpretations, but most workers recognize three main components in the gravity field. The first is a very long (>500) wavelength gravity low which can be interpreted as being due to lithospheric thinning (e.g., Bridwell, 1976; Ander, 1981). The second is an axial gravity high (Decker and Smithson, 1975;

Cordell, 1976, 1978; Ramberg et al., 1978) which in the southern portion of the rift (Fig. 3) has been shown to be primarily the result of crustal thinning (Daggett et al., 1986). Cordell (1982) has shown that this anomaly narrows from south to north, suggesting that extension in the rift also decreases from south to north. The last component of the gravity field is the series of short wavelength lows which are due to the large thickness of fill in rift basins. In the Albuquerque basin, drilling has shown the rift-related fill to exceed 3 km in thickness. In this basin, as well as others, Eocene (Laramide) sediments and Oligocene (pre-rift) volcanics locally form a significant thickness of older basin fill. However, gravity modeling efforts (e.g., Ramberg et al., 1978; Cordell, 1976; Birch, 1982; Wen, 1983; Daggett, 1982; Gilmer et al., 1986; Mattick, 1967) attest to the fact that the Rio Grande rift is associated with an impressive amount of sedimentary fill.

Aeromagnetic Data

An extensive data base of aeromagnetic measurements is available for the Rio Grande rift area. However, these data have not been fully analyzed. Cordell (1983) compiled a digital data base for New Mexico and published a 1:500,000 scale map of the state. Keller et al. (1985) compiled a data base for west Texas and conducted an integrated analysis with gravity data. Himes (1968) studied aeromagnetic data in the northern part of the state of Chihuahua, Mexico. De Angelo and Keller (1988) digitized these data and merged them with data from New Mexico, Arizona, and west Texas. An aeromagnetic map for Colorado has been published (Zietz and Kirby, 1972a,b) but is not in digital form.

Cordell (1976) analyzed aeromagnetic anomalies in the central portion of the rift. He noted relationships which indicated basement control of rift-bounding faults. Anomalies in northern New Mexico were analyzed by Cordell and Keller (1984) and were found to mostly be due to pre-rift structures and volcanics.

Seismic Refraction

Seismic refraction studies provide a general picture of crustal thickness variations in the rift. However, none of the existing studies is very detailed and the width of the rift's crustal structure anomaly can only be estimated from gravity modeling. The northern part of the rift is essentially unstudied by refraction. An interpretive contour map of crustal thickness (Fig. 4) has been constructed using the seismic profiles and gravity anomalies as controls. Upper mantle velocities (Pn) can be estimated to be 7.6 to 7.7 km/s in the rift (Olsen et al., 1979; Sinno et al., 1986) but no truly reversed profile exists along the rift. Existing data in the southern rift have been interpreted to include a 7.4 km/s "rift cushion" (Cook et al., 1979). However, more recent data (Sinno et al., 1986) do not require the presence of such a feature. Additional wide-angle seismic profiles are badly needed throughout the rift.

Seismic Reflection

The only deep seismic reflection profiling experiment in the rift was conducted by the COCORP group (Brown et al., 1979, 1980). These data reveal the basement structure of the Albuquerque basin. A strong reflector at about 7 s correlates with the mid-crustal magma chamber detected through earthquake studies (Sanford et al., 1973, 1977, 1979; Rinehart et al., 1979). A discontinuous reflection at 11-12 s corresponds to the M-discontinuity as mapped by Olsen et al. (1979) from refraction data.

Some industry data have been released for basin studies in the rift. Davis and Stoughton (1979) used seismic reflection data provided by Amoco to determine the deep structure of the San Luis basin in southern Colorado. Keller et al. (1986) used a long reflection profile provided by Exxon to demonstrate the sub-surface extent of Laramide uplifts and basins in the southern New Mexico portion of the rift. Figuers (1987) included a seismic reflection line in his detailed structural analysis of the northern Franklin Mountains. Black (1984) discussed reflection data in the Espanola basin area which depicted the complicated pre-rift structure beneath the basin.

Surface Wave Dispersion

The analysis of surface wave dispersion is a useful but low-resolution way to determine average earth structure along a particular propagation path. Surface waves are most sensitive to variations in shear wave velocity and are thus a val-

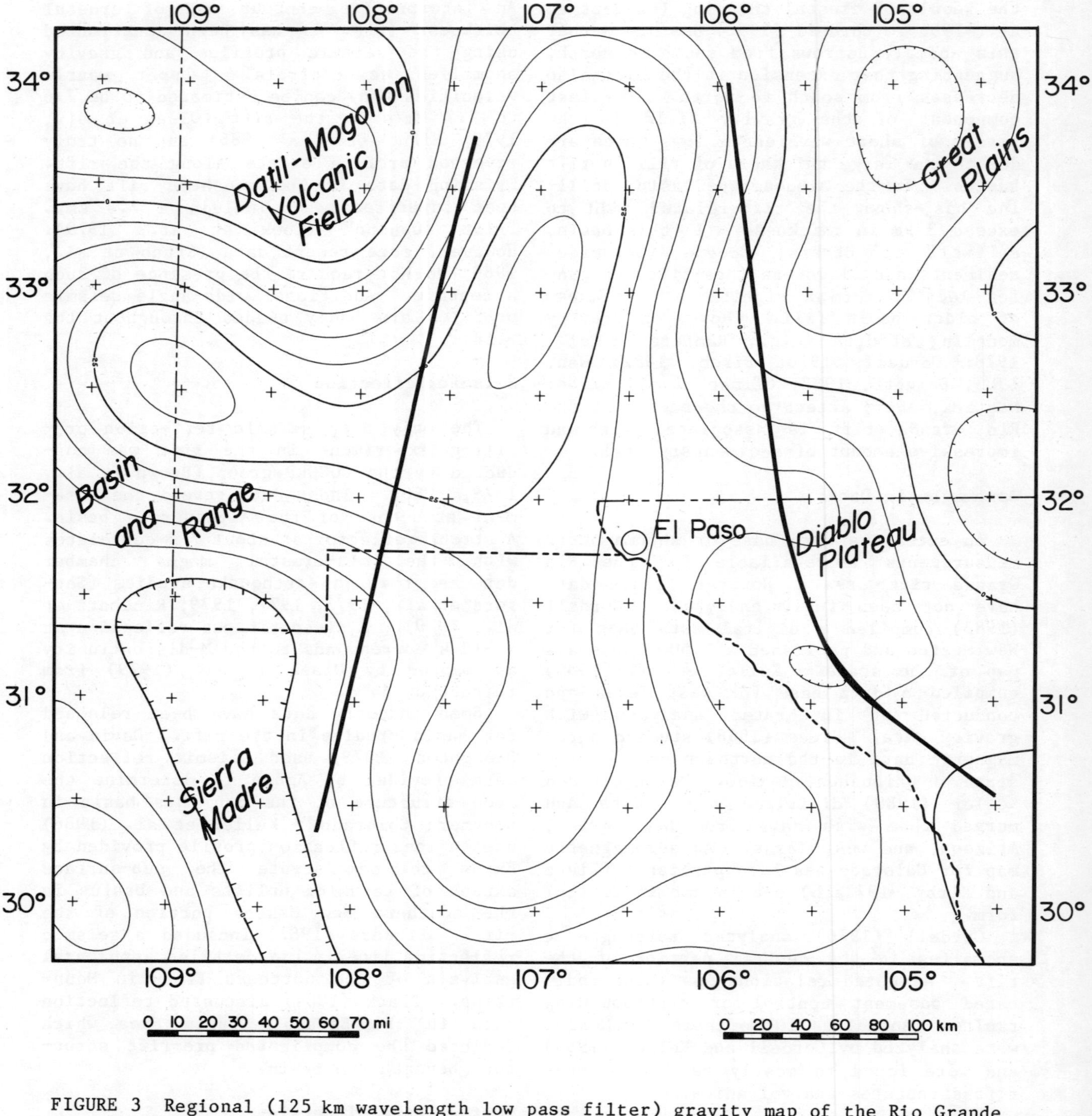

FIGURE 3 Regional (125 km wavelength low pass filter) gravity map of the Rio Grande rift. A variable datum, Bouguer correction was used to isostatically correct the data. Contour interval--5 milligals.

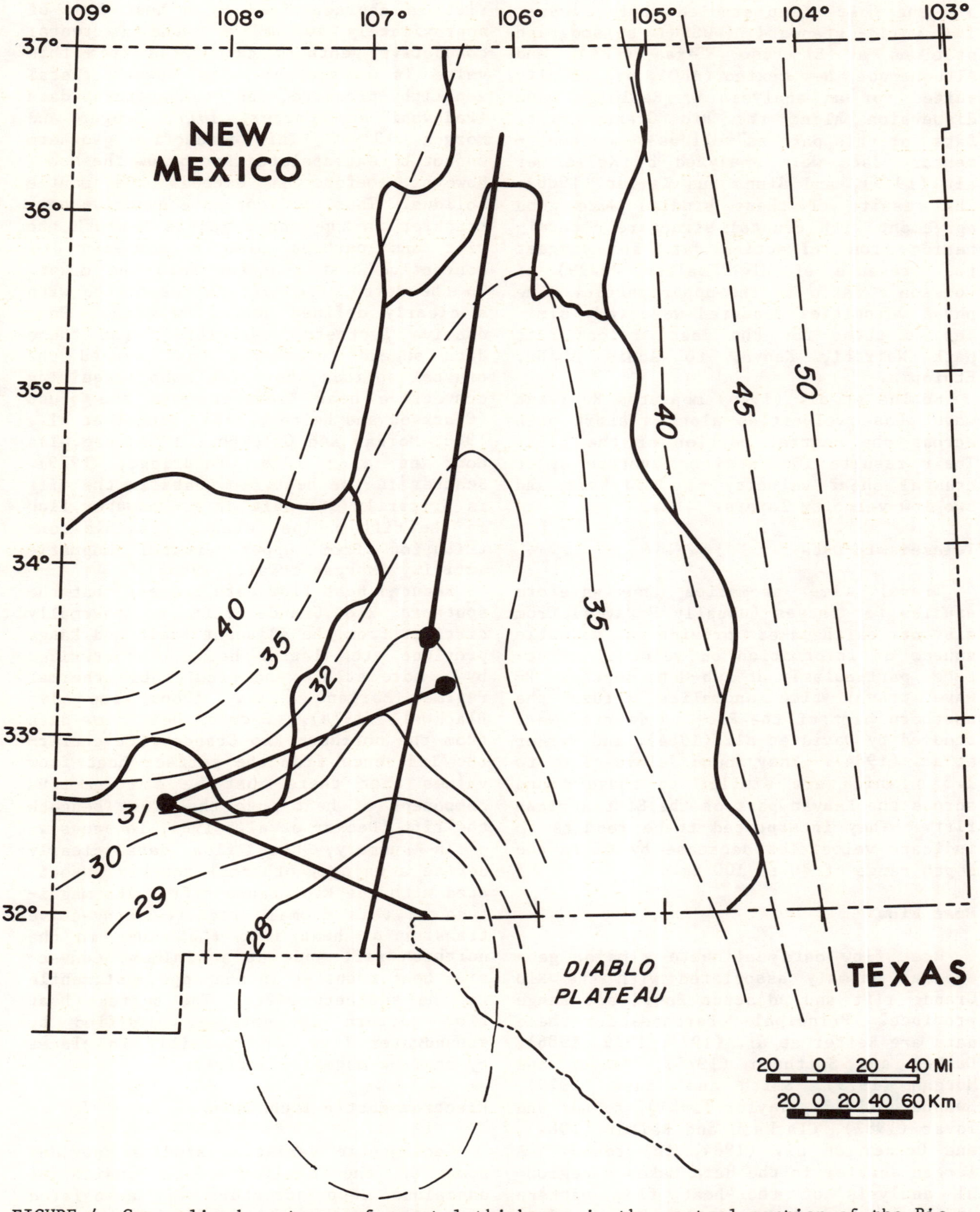

FIGURE 4 Generalized contours of crustal thickness in the central portion of the Rio Grande rift. Dashed contours are less certain. Contour interval varies according to the available control.

uable source of information on this difficult to measure parameter. Rayleigh waves are the most often studied surface wave. The world standard (WWSSN) seismograph stations at El Paso, Texas (EPT), and Albuquerque, New Mexico (ALQ), are ideally suited for an analysis of Rayleigh wave dispersion along the Rio Grande rift. Data for this path as well as some complimentary data were analyzed by Keller et al. (1979), and Sinno and Keller (1986). The results of these studies were good agreement with crustal structure determination from refraction data and suggest the presence of high values (0.29) of Poisson's ratio in the upper mantle. The phase velocities measured were very similar to those for the East African rift path Nairobi, Kenya, to Addis Ababa, Ethiopia.

Schlue et al. (1986) measured Rayleigh wave phase velocities along a short path across the central portion of the rift. Their results indicated an average upper crustal shear velocity of 3.33 km/s and two low-velocity layers.

Teleseismic Data

Travel time anomalies and waveform studies for phases (usually P-waves) from distant earthquakes provide a valuable source of information on velocity structure, particularly at sub-Moho depths. P-wave travel time anomalies across the northern part of the Rio Grande rift were studied by Davis et al. (1984), and Parker et al. (1984). They found delays of up to 1.5 s which are similar to those found across the Kenyan part of the East African rift. They interpreted these results to indicate velocities decrease by 8% in the depth range of 70 to 100 km.

Heat Flow

Heat flow data delineate a major geothermal anomaly associated with the Rio Grande rift and adjacent Basin and Range province. Principal references for these data are Reiter et al. (1974, 1979, 1986), Decker and Smithson (1975), Seager and Morgan (1979), Smith and Jones (1979), Swanberg (1979), Taylor (1981), Reiter and Tovar (1982), Clarkson and Reiter (1984), and Decker et al. (1984, in press). A larger scatter in the data makes a regional analysis of the heat flow pattern difficult, and contour maps of these data are somewhat subjective (e.g., compare Reiter et al., 1975; Swanberg, 1979; Sass et al.,1981). Decker and Smithson (1975) suggested that the southern Rio Grande rift is characterized by a heat flow of approximately 100 mW m^{-2} and a crustal conductive geotherm calculated from this value is consistent with lower crustal xenolith pressure and temperature data (Padovani and Carter, 1977; Seager and Morgan, 1979). This conductive geotherm cannot be extrapolated far below the Moho, however, before it exceeds the mantle solidus. Thus, indicating convective heat transfer in the upper mantle beneath the rift and perhaps also a component of convective heat transfer into the crust. To the north, the rift is associated with a clearly defined heat flow high. Conductive geotherms calculated from these data suggest geotherms that exceed the crustal solidus above the Moho, requiring convective heat transfer into the crust (Clarkson and Reiter, 1984; Decker et al., 1986; Morgan and Golombek, 1984; see also Cook et al., 1978; Baldridge, 1979). Scatter in the heat flow data in the rift is primarily a result of groundwater flow in the rift basins, with a possible contribution from upper crustal magmatic activity (Morgan et al., 1986a).

Reduced heat flow data suggest that the southern Rio Grande rift is thermally distinct from the adjacent Basin and Range province, the latter being characterized by a more stable and predictable thermal regime (Morgan et al., 1986a; see also Blackwell, 1978). Reduced heat flow data from the northern Rio Grande rift confirm the inference from the surface heat flow values that there must be a convective component of heat into the crust beneath the rift (Decker et al., 1984, in press).

In summary, heat flow data clearly define a major geothermal anomaly associated with the Rio Grande rift. The magnitude of this anomaly requires convective transfer of heat into the crust in the northern rift and, at a minimum, convective heat transfer in the uppermost mantle in the southern rift. The surface heat flow pattern is severely modified by groundwater flow, and possibly in places by shallow magmatic activity.

Electromagnetic Techniques

Geomagnetic variation studies provided some of the earliest data indicating anomalous deep structure was associated with the Rio Grande rift (Schmucker, 1964). More recent magnetotelluric measurements (e.g., Jiracek et al., 1979,

1983, 1987; Hermance and Pederson, 1980) have provided information about the conductivity of the crust that suggest many anomalous features are present.

Seismicity

The Rio Grande rift region has been known to be seismically active for many years (e.g., Reid, 1911). Instrumental studies began in 1960, and enough data had been gathered primarily in New Mexico to warrant review articles by Sanford et al. (1972, 1979). An updated seismicity map of New Mexico has recently been compiled by Stover et al. (1988). The most active area in this century has been between Socorro and Albuquerque, New Mexico. The largest events in this area occurred during a swarm which extended from July 1906 to January 1907. Two of these events reached intensity VIII. Studies of the seismicity in this area suggested the presence of a magma chamber (Sanford et al., 1973, 1977, 1979; Rinehart et al., 1979) at mid-crustal depths which correlated with an amplitude anomaly in COCORP (Brown et al., 1979, 1980) and refraction (Olsen et al., 1979) data. Recent earthquakes in the Albuquerque, New Mexico, area were studied by Jaksha and Sanford (1986). Their focal mechanisms indicated WNW-ESE extension.

Seismicity in the northern New Mexico portion of the rift came under investigation through studies by Los Alamos National Laboratory beginning in 1972. The western boundary of the rift near Espanola was found to be particularly active (Sanford et al., 1979; Cash and Wolff, 1984).

Seismicity data in the southern portion of the rift are scarce. A seismograph station has been operating in El Paso since the early 1960s and was upgraded to WWSSN status in 1977. Seismicity along the easternmost portion of the rift in West Texas was monitored between 1977 and 1981 and discussed by Dumas et al. (1980) and Dumas (1981). A large ($M = 6.4$) earthquake occurred in this area in 1931 (Byerly, 1934a,b). Doser (1987) conducted a complete study of this event and suggested it occurred along the western boundary of a graben which can be considered the easternmost rift structure in the area.

Another very large event occurred in the northeasternmost portion of the State of Sonora, Mexico, near Douglas, Arizona, in 1887 (Aguilera, 1920; Sbar and DuBois, 1984). This event created a fault scarp 76 km long with a maximum height of 4.5-5.1 m. Its location is about 100 km west of the accepted western boundary of the rift, but at this latitude the rift is widening and becoming diffuse, much as the Kenyan portion of the East African rift widens as it passes southward into Tanzania.

These seismic studies along with studies of Quaternary fault scarps (e.g., Muehlberger et al., 1978; McCalpin, 1982, 1987; Coleman et al., 1985; Machette, 1986, 1987a, b; Menges, 1987b; Foley et al., 1988) attest to the continuing tectonic activity of the rift. In general, this activity is moderate and characterized by swarms of events. Focal mechanisms generally indicate E-W extension, although there are many local variations (Sanford et al., 1979; Doser, 1987).

Recent Crustal Movements

Ongoing tectonic activity in the Rio has been investigated through the analysis of leveling data. The Socorro, New Mexico, area has been of particular interest (e.g., Reilinger and Oliver, 1976; Reilinger et al., 1980a; Larsen et al., 1983, 1986). The most recent results (Larsen et al., 1986) indicate an average uplift rate of 0.23 cm/yr from 1912 to 1951 and of 0.18 cm/hr from 1951 to 1980. Geomorphic evidence suggests an uplift rate of 0.18 cm/yr over the past 20,000 years (Ouchi, 1983). This uplift is attributed to the inflation of the mid-crustal magma chamber in the area (Reilinger and Oliver, 1976).

Savage et al. (1980) measured horizontal strain in the area and found a slight horizontal contraction. Larsen et al. (1986) showed that this contraction as well as the observed areas of subsidence could be modeled with deflating spherical sources at 19 or 35 km feeding an inflating sill at 19 km.

Significant vertical movements have also been observed in West Texas (Reilinger et al., 1980a; Ni et al., 1981) in areas which are probably associated with the rift.

Stress Measurements

Studies of paleostress and contemporary stress indicators (Zoback and Zoback, 1980) indicate that the middle Tertiary was a time of ENE-WSW extension in the western U.S. The stress direction rotated about 45° to WNW-ESE extension in the late

Tertiary (e.g., Zoback et al., 1981). Although there are local complications, this pattern generally holds true for the Rio Grande rift area (Aldrich et al., 1986; Henry and Price, 1986). Focal mechanism studies (e.g., Sanford et al., 1979; Jaksha and Sanford, 1986; Doser, 1987) indicate strike-slip movements are important locally along the rift.

DESCRIPTIONS OF FIELD TRIP STOPS

As organized, this field trip will proceed from El Paso, Texas, up the Rio Grande rift to central Colorado (Fig. 5). Brief descriptions of each stop follow, but in most cases the reader is referred to descriptions in other guidebooks, particularly New Mexico Bureau of Mines and Mineral Resources Circular 163, which was prepared for the 1978 International Rift Conference in Santa Fe. This trip should not be undertaken without a copy of Circular 163 which contains formal road logs and much supplementary material. The numbers in parentheses following the stopnumber refer to stop numbers in Circular 163. The New Mexico Geological Society is a very active organization which sponsors an annual field conference with an accompanying guidebook. These publications contain a wealth of information about the Rio Grande rift which is often overlooked in the international literature.

Stop 1--Tom Lea Park

This stop overlooks the cities of El Paso and Juarez and provides an excellent opportunity to discuss the geologic setting of the region. The Sierra de Juarez lie behind the city of Juarez and are composed primarily of folded and thrusted Cretaceous strata. A complete sequence of Paleozoic strata are exposed in the Franklin Mountains which lie to the northeast (Circular 163, p. 60; LeMone and Cornell [1988], p. 7).

Stop 2--Fault Scarp

The Franklin Mountains are bounded on the east by a series of normal faults at least one of which has experienced Quaternary movement. A well-developed scarp parallels Magnetic Street at this location (Circular 163, p. 61; LeMone and Cornell [1988], p. 13).

Stop 3 (Sla)--Wilderness Park Museum and Precambrian Outcrops

The Wilderness Park Museum provides information on the flora and fauna of the area. The Franklin Mountains are, in fact, a very large state park. A series of roadcuts above this museum provide excellent exposures of the Precambrian rocks present (Circular 163, p. 63; LeMone and Cornell [1988], p. 13-15).

Stop 4--Smuggler's Pass

This pass through the Franklin Mountains was reportedly used by smugglers for many years. From this location, one has an excellent view of the Rio Grande rift. The Franklin Mountains are in fact an intrarift horst. To the east lie the Hueco bolson and Hueco Mountains, and to the west lie the Mesilla Bolson and Portrillo Mountains (Circular 163, p. 64; LeMone and Cornell [1988], p. 15-16).

Stop 5 (Slb)--Tom Mays Park

The overlook in this park provides a beautiful view of the Rio Grande Valley, Mesilla Bolson, and the western portion of the rift (Circular 163, p. 65-70; LeMone and Cornell [1988], p. 16-22). Two recent maps (Seager et al., 1982, 1987), with geologic sections and gravity profiles across major range and basin features, cover the south-central New Mexico region between this stop and Caballo Reservoir (Stop 11). Hydrogeologic sections of the Mesilla and southern Jornada basins (Hawley, 1984) provide a more detailed interpretation of basin structure and upper Cenozoic basin-fill stratigraphy in the area covered by Stops 5 to 9. Subsurface studies of the basin fill sequence (Santa Fe Group) by Hawley and Lozinsky (unpublished) indicate that the maximum thickness of Miocene to mid-Pleistocene deposits in the Mesilla Bolson is about 1,000 m. The time of major offsets along basin-bounding and intra-basin faults and most rapid aggradation of central basin grabens was probably during the middle to late Miocene (~ 12 to 5 ma).

Stop 6--Sierra Cristo Rey and Mesilla Valley Fault

Facing downstream, the Rio Grande here flows through the historic "Paso del Norte del Rio Bravo." Limestone cliffs on the east and west banks are the Finlay Lime-

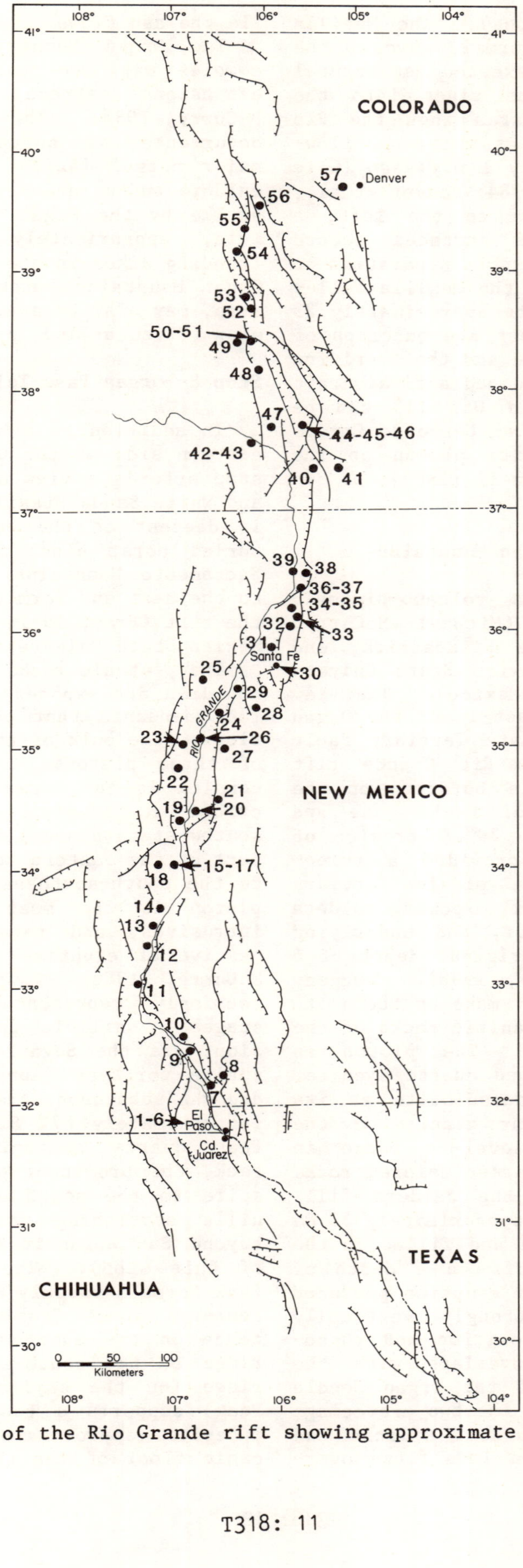

FIGURE 5 Index map of the Rio Grande rift showing approximate locations of field trip stops.

stone (Albian, Cretaceous). The Mesilla Valley Fault zone, a normal, down to the west, fault system exerted structural control and guided the river into the Lower Mesilla Valley area when the Rio Grande was established as a through-flowing stream approximately 1 m.y. ago (Gile et al., 1981; Hawley, 1984; Seager et al., 1987). In the cliffs to the left, a number of slickenside surfaces record fault traces. Stratigraphic separation on this possible splay of the Mesilla Valley fault zone appears to be approximately 25 m. Directly to the west are outcrops of the Mesilla Valley shale and the overlying Anapra sandstone. Abandoned anti-aircraft gun emplacements (World War II) can be found in the foothills of Cerro de Cristo Rey, the Eocene andesite pluton on the skyline (LeMone and Cornell [1988], p. 36-37).

Stop 7--Overview of Organ Mountains

The Organ Mountains volcano-plutonic complex, New Mexico (Michael McCurry, Donald Butcher, Delane Fitzpatrick, and Steven Yanicak, New Mexico State University, Las Cruces, New Mexico). The view to the east and southeast is of the Organ Mountains, a typical late Tertiary fault block in the southern Rio Grande rift (Seager, 1981). It is bordered on the east by a major normal fault zone and tilted westward 15° to 20°. Erosion of the fault block has provided a three-dimensional view of a middle Tertiary volcano-plutonic complex, exposing caldera fill, the caldera floor, and underlying plutonic rocks to an original depth of 6 km (Seager and McCurry, 1988). Rugged, light-colored peaks that make up the north end of the range are granitic rocks of the Organ Needle pluton. The pluton is dominantly coarse grained quartz syenite, and it varies from monzodiorite at its core to alkali feldspar granite at the structurally highest level--a range in SiO_2 of 55% to 75%. Darker colored rocks to the southeast are the caldera fill. Two major eruptions at approximately 33 Ma caused caldera collapse and filled in the caldera with welded tuff to a depth of about 3 km. The second eruption produced deposits that are strongly vertically zoned in chemical composition and phenocryst content. It overlaps with the compositional zoning of the Organ Needle pluton, suggesting that the two are comagmatic. Several postcaldera silicic and intermediate composition lava flows overlie the ash flows.

The Organ Mountains volcano-plutonic complex may have evolved in a regional extensional stress field (Seager and McCurry, 1988). This is suggested by the occurrence of north-northwest trending major normal faults that cut the intra-caldera tuffs, and are also intruded along strike by the Organ Needle pluton. Silicic, approximately 33 Ma, northwest trending dikes in the northern part of the Organ Mountains batholith, near the next stop, may also be a product of such extension (Circular 163, p. 73-74).

Stop 8--Organ Pass-Tularosa Basin

In addition to a beautiful view of the eastern side of the Organ Mountains, this stop affords a view of the Tularosa basin and White Sands Missile Range. This basin is deepest on the west, and there is a buried horst block in its center. The Sacramento Mountains lie in the distance to the east and form the eastern margin of the rift (Pray, 1961; McCurry, et al., New Mexico State University).

The plutonic rocks underlying the Organ cauldron are exposed to the south, as are roof pendants that comprised the caldera floor. The bulk of the batholith consists of three plutons. The oldest and most complex is the Organ Needle pluton (described at previous stop), forming the southwestern pinnacled skyline and a large part of the eastern bulge of the batholith to the south-southeast. The Organ Needle pluton is the most voluminous of the intrusives, and retains structures suggestive of eruptive drawdown (Seager and McCurry, 1988). The younger plutons, seemingly representative of successive stages of silicic interior magma intrusion, are the Sugarloaf Peak quartz monzonite porphyry (approximately 62% SiO_2) and the subsequent granite of Granite Peak (approximately 67% SiO_2). The Sugarloaf Peak quartz monzonite forms Sugarloaf Peak, the prominent gray-white exfoliated spire to the south and the pediment and hills stretching northward towards and beyond San Augustin Peak (one mile north of this stop). The granite of Granite Peak forms a roughly cylindrical intrusion centered about Granite Peak, the prominence on the east end of the high flat ridge to the south-southeast. The same ridge (on the skyline, beyond Sugarloaf Peak) supports a roof pendant of Orejon andesite (approximately 38 Ma), the volcanic floor of the cauldron. Other pre-cauldron country rocks are precambrian granites, diorites, amphibolites and schists, and paleozoic limestones and siliciclastics.

Near the town of Organ to the west, Laramide structures and batholith associated porphyry and skarn mineralization are exposed (Seager, 1981).

Stop 9 (S3)--Leasburg Dam-Radium Springs

There is a discussion of this stop in Circular 163 (p. 75-78).

Stop 10 (S4)--Rincon Overlook

There is a lengthy discussion of this stop in Circular 163 (Circular 163, p. 79-81).

Stop 11 (S6)--Caballo Reservoir

There is a discussion of this stop in Circular 163 (p. 85-87). This locality is at the northern edge of the region mapped by Seager et al. (1982, 1987). See the 1986 New Mexico Geological Society Guidebook for details on the geology of the Truth or Consequences area north of this stop. Recent seismotectonic studies by the U.S. Bureau of Reclamation and the Geological Survey (Foley et al., 1987; Machette, 1987b) document a significant amount of Late Quaternary offset along segments of the Cabllo fault between Caballo Dam and Truth or Consequences. The fault has been assigned a maximum credible earthquake (MCE) magnitude of Ms 7-1/4 by the Bureau of Reclamation.

Stop 12 (S7ab)--Elephant Butte Reservoir and Dam

There is a discussion of this stop in Circular 163 (p. 89-90).

Geology of the Elephant Butte area, south-central New Mexico (Richard P. Lozinsky, Leighton and Associates, Walnut, California). The Elephant Butte area lies within two structural subdivisions of the Rio Grande rift, the Engle basin, and the Cutter Sag. The boundary between these two structural features is marked by the mostly submerged Hot Springs fault. This fault dips about 70-80° to the west and has at least 1,220 m of down-to-the-west separation (Lozinsky, 1986). Although most of the movement on the Hot Springs fault has been dip-slip, there is evidence for strike-slip motion early in its history. To the north and south, the Hot Springs fault merges with other faults to become the range-bounding structure for the Fra Cristobal Range and the Caballo Mountains, respectively. Some faults in the area show evidence for recent movement (Machette, 1987).

The eastward-tilted Engle basin covers an area of about 1,200 km^2 and is filled with more than 700 m of late Oligocene to middle Pleistocene Santa Fe Group basin-fill sediments (Lozinsky, 1987). Most basin-fill exposures in the Engle basin are sands and gravels of the Plio-Pleistocene Palomas Formation (upper Santa Fe Group; Lozinsky and Hawley, 1986). Two facies, the piedmont and axial-river, comprise the Palomas Formation. The piedmont facies consist of alluvial-fan deposits derived from the surrounding uplifts, whereas the axial-river facies were deposited by the ancestral Rio Grande. The sandy beaches along the western shore of the reservoir occur in axial-river deposits. Basalt flows and maar deposits that range in age from 2-3 Ma are interbedded in the Palomas Formation. The extensive surface on the west side of the Rio Grande is the Cuchillo. It is a constructional surface and represents the highest aggradational level of the Engle basin at the end of Palomas Formation deposition (Lozinsky and Hawley, 1986). Just to the south, the northeastward-tilted Mud Springs uplift separates the Engle basin from the Palomas basin to the south.

The Cutter Sag is the structurally low part of the Caballo-Fra Cristobal uplift. Rocks exposed here consist mainly of the Upper Cretaceous Crevasse Canyon Formation (upper Mesaverde Group) and Upper Cretaceous-Eocene (?) McRae Formation. Numerous plant and dinosaur fossils have been recovered from the lower part of the McRae Formation (Wolberg et al., 1986). Most high mesas in this region are capped by 2-3 Ma basalt flows. A dike, with mantle-derived inclusions, which cuts the Mesaverde Group at Stop 12, and the plug that forms Elephant Butte (and intrudes the McRae Formation) are exposed parts of the conduit system that fed these flows. Also see Baldridge (1979) and Warren et al. (1979) for detailed information on the geochemistry and petrography of inclusions in Pliocene basalts of the Engle basin.

One of the more interesting features of the Elephant Butte area is the now-submerged course of the Rio Grande. This is one of the few areas along the Rio Grande where the river has crossed the basin-bounding fault to cut a channel into the more resistant bedrock of the margin uplift. Lozinsky (1986) presents a model involving stream piracy and capture to explain how this phenomenon developed.

Stop 13 (S8)--Fra Cristobal Range

There is a discussion of this stop in Circular 163 (p. 91-96).

Stop 14 (S9)--San Marcial Basin, San Mateo Mountains

There is a discussion of this stop in Circular 163 (p. 96-98).

Stop 15--Socorro, New Mexico

Socorro is the home of the New Mexico Institute of Mining and Technology. The 1983 New Mexico Geological Society Guidebook and a recent field guide to the Socorro area by McLemore and Bowie (1987) cover the region in detail. From a geophysical point of view, the area is underlain by a mid-crustal magma chamber which has been detected by a variety of seismic data (Sanford et al., 1973, 1977, 1979; Circular 163, p. 115-125).

Stop 16 (S10)--Overview of Socorro Area

As shown in Figure 6, the north-trending basins and ranges of the Socorro region (including Socorro Peak) are predominantly tilted fault blocks (half basin, half uplift) formed by westward extension of the continental crust along the Rio Grande rift. Twentieth century uplift, historic seismicity, thermal springs, and high-heat flow in the Socorro Peak-Socorro Basin are reasonably attributed to an active sill-like magma body, about 18 km below the Rio Grande valley, and numerous dike-like bodies about 4-14 km below Socorro Peak (Sanford, 1983).

The 2,000-ft escarpment forming the east face of Socorro Peak represents a late-stage tilted-fault-block uplift within the Rio Grande rift. Less than 7 Ma old, the Socorro-Lemitar uplift has disrupted an early closed basin of the rift known as the Popotosa Basin. Silicic lava flows and domes, that now form the skyline west of "M" Mountain, were erupted episodically onto the playa floor of the Popotosa Basin from 12 to 7 Ma. Claystones of the upper Popotosa Formation lack sufficient strength to support the flanks of the uplifted lava flows. The Socorro Mountains are therefore surrounded by a hummocky apron of Pleistocene landslide blocks compositionally equivalent to lava flows found immediately upslope.

The bold cliffs on the east face of Socorro Peak (below the "M") are formed by extremely well indurated debris-flow deposits of the lower Popotosa Formation. These coarse matrix-supported clastics occur as thick wedges in narrow strike valleys (5 km wide) adjacent to low-angle early rift faults. A widespread angular unconformity at the base of the Popotosa (20-30°) and abrupt thickness variations in underlying ash-flow tuff sheets indicate that domino-style crustal extension began along the rift axis about 29 Ma.

Thermal springs flowing from the east flank of the Socorro Mountains lie at the intersection of the range-bounding fault zone and the Socorro transverse shear zone--a broad domain boundary separating westerly tilted fault blocks on the north from easterly tilted fault blocks on the south (Fig. 6).

The relatively young Socorro Mountain block transects the northeastern topographic wall and structural margin of the 32 Ma old Socorro cauldron (Fig. 6). The uplift thus provides a cross-section-like view of the cauldron margin. Pennsylvanian limestones (Madera Formation) in the middle slope of Socorro Peak define the structurally high rim of the cauldron; and a large rhyolite lava dome in the caldera moat (north of Blue Canyon) is considered to mark a point on the ring-fracture zone. Moat deposits, collectively assigned to the Luis Lopez Formation, thin abruptly by at least 200 m northward across the buried rim of the Socorro cauldron.

The middle Tertiary caldera complex southwest of Socorro and the area of strong domino-style extension along the axis of the rift are superimposed on a late Laramide (Eocene) transpressional welt locally bound by right-lateral wrench faults (Fig. 6). The Socorro region can be considered a relatively soft (warm) zone in the continental lithosphere that has been periodically squeezed, transposed, intruded, and stretched between two relatively rigid (cold) microplates, represented by the Colorado Plateau and the Great Plains provinces (Fig. 6). More details on the geology and tectonic setting of this area are available in Chapin et al. (1978), Chamberlin (1980, 1981, 1983), Chamberlin and Osburn (1986), and

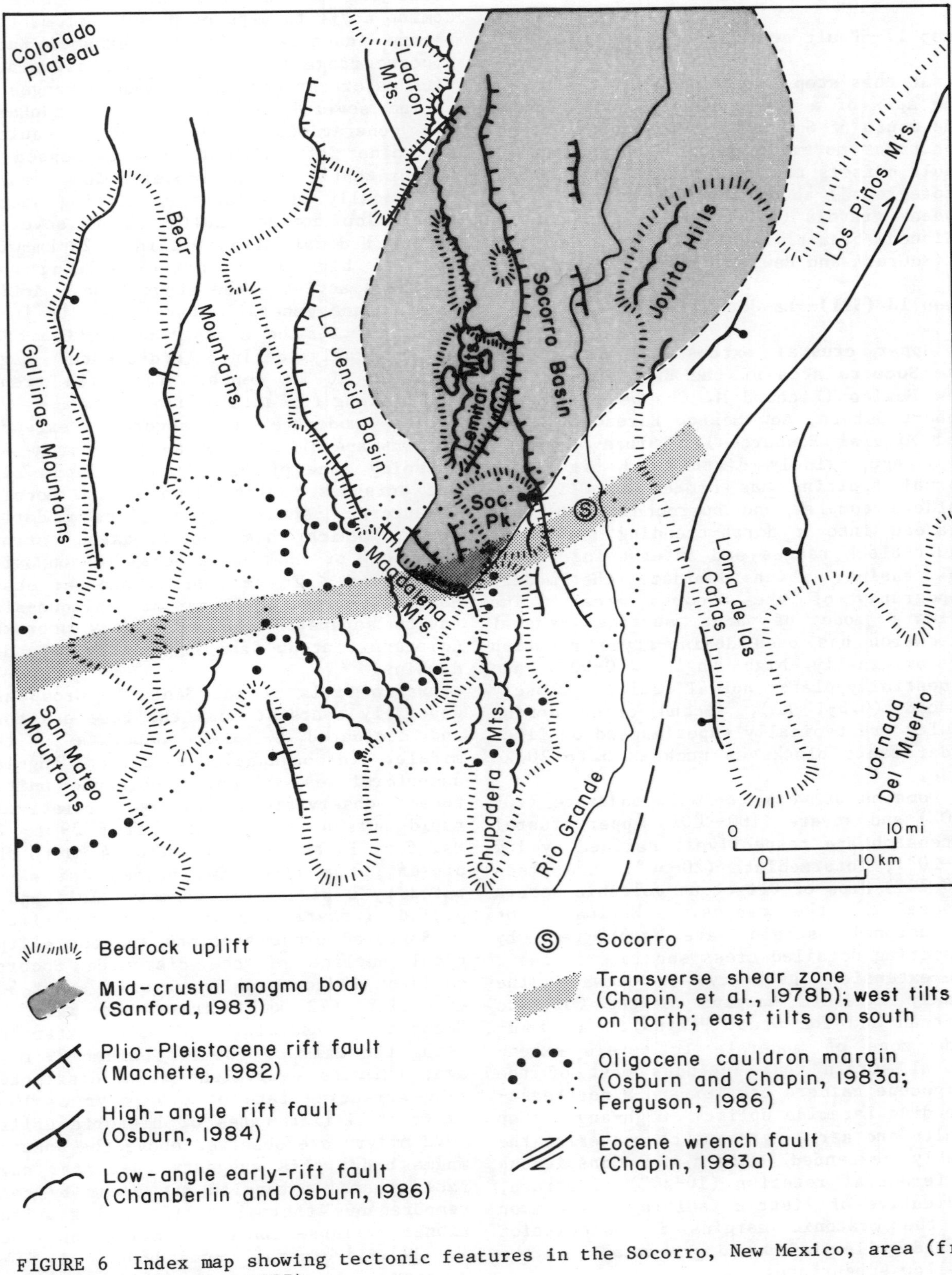

FIGURE 6 Index map showing tectonic features in the Socorro, New Mexico, area (from McLemore and Bowie, 1987).

McLemore and Bowie (1987) (Circular 163, p. 121-128).

Stop 17—Fault scarp

At this stop, a piedmont fault scarp at the apex of a Pleistocene alluvial fan at the mouth of Socorro Canyon can be viewed. Recurrent movement on this range-bounding fault offsets the late Pleistocene terrace (foreground) about 3 m, the middle Pleistocene gravels (9:00) about 30 m, and the Pliocene basalt flow (3:00) by at least 60 m (McGrath and Hawley, 1987, Fig. 1).

Stop 18 (S11)—La Jencia Basin

Upper crustal extensional domains in the Socorro area of the Rio Grande rift, New Mexico (Richard M. Chamberlin and G. Robert Osburn, New Mexico Bureau of Mines and Mineral Resources). Since about 29 m.y. ago, widely distributed high-angle normal faulting has broken an Oligocene caldera complex and surrounding volcanic plateau into a north-trending series of fault-block ranges and intervening alluvial basins, which dominate the modern topography of the Socorro area. The primary mode of near surface crustal extension has been domino-style rotation of originally high angle (60-90°) and essentially planar normal faults. Closely spaced (0.5-1 km), second order domino faults are typically superimposed on first order fault blocks as much as 5 to 20 km wide.

Domains of weak (2-20%), moderate (20-100%) and severe (100-200%) upper crustal extension are respectively defined by low (0-20°), intermediate (20-50°), and steep (50-70°) dips of Oligocene volcanic strata exposed in the ranges. Estimates of extensional strain are supported by restoring detailed cross sections to their pre-extension geometry. Between the weakly extended margins of the Colorado Plateau and the Great Plains is a 45-km-wide zone of moderate to severe extensional strain that includes most of the Oligocene caldera complex and a northeast-trending Laramide uplift. High-angle step faults and narrow grabens characterize the weakly extended domain. Conspicuous differential rotation (10-30°) of strata, indicative of listric faulting, is common on the cratonic margins in association with ductile shales and gypsum beds in the shallow subsurface.

The moderately extended domain is associated with first order fault blocks about 5 km wide and second order blocks about 0.5 km wide that have been rotated domino style to dips of 30-40° before they became inactive. Several exhumed Early Miocene range front faults are associated with major canyons in the modern ranges

The severely extended domain contains two generations of domino-style faults, the older of which is now expressed as subhorizontal younger-over-older faults cut locally by moderately dipping second generation domino faults. The severely extended domain lies within a 25 km-wide belt of high-angle normal faulting that has been active since about 5 Ma. Active fault zones generally parallel, dip into, or transect, the surface projection of a geophysically defined mid-crustal magma body at 20 km depth, and which trends N20°E along the axis of the rift.

The moderate to severely extended domains are bisected by an east-northeast-trending zone of high-angle oblique slip and scissors faults known as the Socorro transverse shear zone (accommodation zone), which has accommodated strong rotation of domino blocks in opposite directions. To the north, another east-northeast-trending zone of oblique-slip faults separates the relatively unbroken Colorado Plateau from the highly extended domains.

Basin fills of the Santa Fe Group are generally coarsest near the base and top, and dominated by playa deposits in the middle. In conjunction with local angular unconformities in dated volcanic units, these observations suggest relatively rapid pulses of extension from 29 to 20 Ma, from 16 to 9 Ma, and from 4 Ma to the present; the first two pulses are separated by a widespread magmatic lull and a period of thermal subsidence in the rift.

Restored cross sections and the elliptical outline of the distended Socorro cauldron indicate an average of about 50% extension (22 km) across the rift at Socorro. Assuming episodic extension along the axis of the rift since 29 m.y. ago, this is equivalent to a distinctly slow spreading rate of 0.9 mm/yr or 10^{-1} cm/yr. Similar rates of historic uplift (0.2 mm/yr) are observed above the Socorro magma body; this suggests that near surface crustal extension may occur by contemporaneous thermal uplift and gravitational collapse above a ductile zone such as the mid-crustal magma body. A 5 km spacing of first order fault blocks in the Early Miocene implies a much shallower zone of decoupling at this time. Thus it seems that domains of active crustal extension have shifted in space and time in response to waxing and waning thermal regimes of local and regional extent. See Machette (1986), McGrath and Hawley (1987), and Arizona Geological Society Digest (v. 16, p. 464) for additional discussion (Circular 163, p. 128-130).

Stop 19 (S12)--Rio Salado Rest Area

There is a discussion of this stop in Circular 163 (p. 132-137).

Stop 20 (S13)--Black Butte

There is a discussion of this stop in Circular 163 (p. 138-139).

Stop 21--Abo Pass, Albuquerque Basin

This stop provides an overview of the Albuquerque basin and of the region traversed by the COCORP profile in the area. Shell Oil has had an active exploration program in this basin. This area is at the southern edge of the Albuquerque Basin region covered by the 1982 New Mexico Geological Society Guidebook (Day 1, Stops 4 to 6).

Structure and stratigraphy of the Albuquerque basin (Richard P. Lozinsky, Leighton and Associates, Walnut, California). The Albuquerque basin, covering an area of about 7,400 km^2, is one of the largest basins of the Rio Grande rift (Lozinsky, 1988). The basin is well defined on the eastern margin by the Sandia-Manzano-Los Pinos uplift; however, it is less well-defined by topographic features along the western margin. Seismic reflection work by the Shell Oil Company (staff communication, December 1986) shows that the basin consists of a northern, eastward-tilted half-graben and a southern, westward-tilted half-graben. The transition zone between these two half-grabens appears to occur along a southwest extension of the Tijeras fault.

Basin structure generally consists of a deep, inner graben that is flanked by relatively shallow ramps leading up to the margin areas (Fig. 7). These faulted ramps form a step-down progression into the deep, inner basin. Faults that greatly displace the basin fill occur within the basin and are not near the topographically high margins. The vertical distance between the Precambrian strata on top of the eastern margin uplifts and the deepest part of the basin indicates that the structural relief is well over 10,000 m.

The major fault in the north half-graben is located approximately under the present-day Rio Grande as evidenced by the basin fill thickness difference between the Transocean Isleta #1 well and the Shell Isleta #2 well. The basin fill thickness increases from 1,536 m in the Transocean well to 4,407 m in the Isleta #2 well over a distance of about 8 km (Lozinsky, 1988). A major displacement fault (or faults) with as much as 2,871 m of offset is postulated between the two wells. Based on seismic data, the major throw fault in the south half-graben appears to be located under the present position of the Rio Puerco. Many of these faults in the Albuquerque basin appear to be listric according to seismic data.

The major basin-fill unit of the Albuquerque basin is the latest Oligocene to middle Pleistocene Santa Fe Group. Within the Albuquerque basin, the thickness of the group ranges from about 1,000 to 2,000 m along the margins to as much as 4,407 m in the central part of the basin. This thick section was penetrated in the Shell Isleta #2 well and represents the thickest documented section of the Santa Fe Group in the entire Rio Grande rift (Lozinsky, 1988).

Pre-Santa Fe Tertiary deposits underlie the Santa Fe Group and indicate the existence of two structural basins that predate the Albuquerque basin. These deposits are subdivided into the Galisteo-Baca formations and an overlying, informal unit, the unit of Isleta #2 well. This informal unit is at least partly equivalent to the Datil Group (as defined by Osburn and Chapin, 1983) and the Oligocene volcanic sequence that overlies the Datil Group. Thicknesses are up to 454 m in the Galisteo-Baca deposits and as much as 2,185 m in the unit of Isleta #2 well (Lozinsky, 1988). Upper Cretaceous strata underlie the basin fill deposits in most regions of the basin, except along the eastern margin where the Santa Fe Group rests on pre-Cretaceous strata.

Stop 22 (S14)--Cerro de los Lunas

There is a discussion of this stop in Circular 163 (p. 141-143). Also see 1982 New Mexico Geological Society Guidebook (Day 2, Stop 1), Baldridge et al. (1987), and Lozinsky (1988).

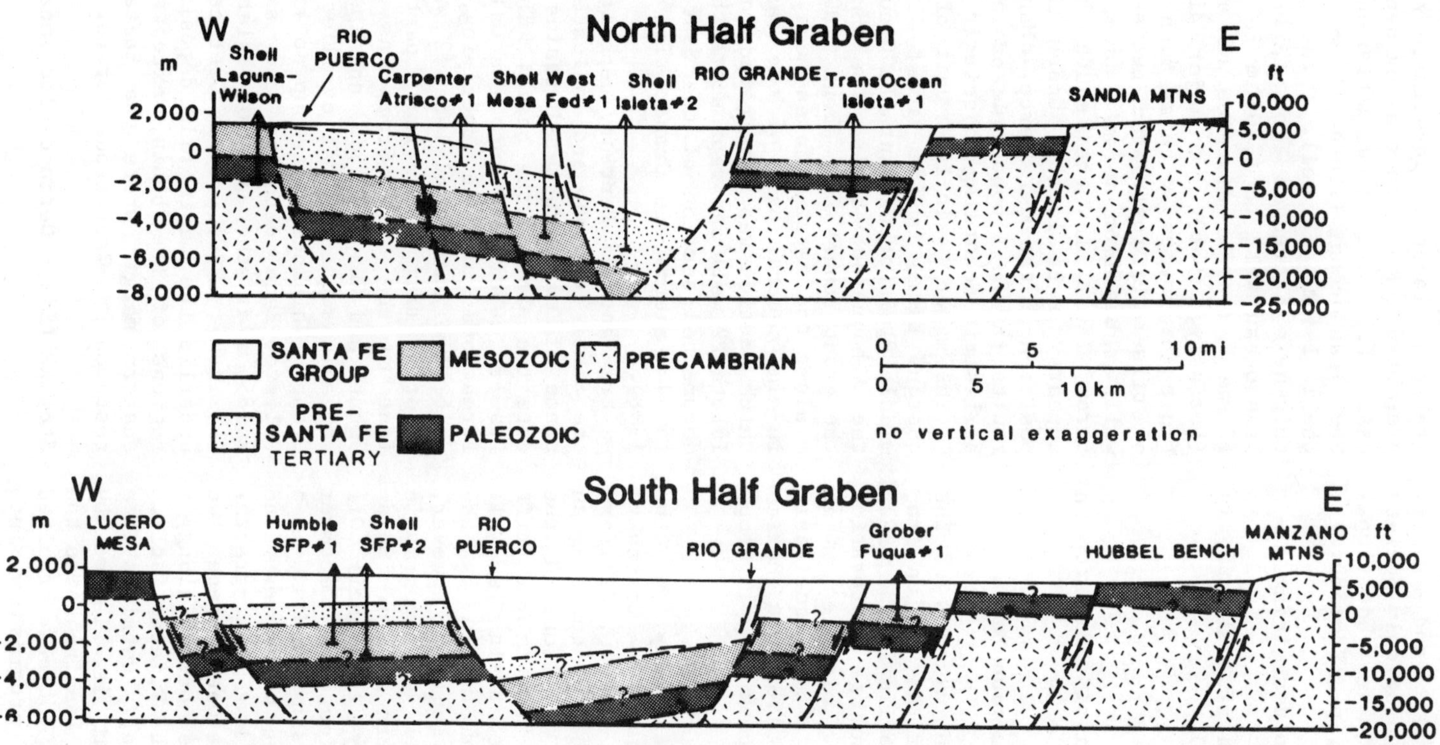

FIGURE 7 Cross-sections of the Albuquerque basin. The northern cross-section is for the portion of the basin near the city of Albuquerque. The southern cross-section is for the portion of the basin near Belen.

Stop 23 (S16)--Albuquerque Overlook

There is a discussion of this stop in Circular 163 (p. 146-158). Also see 1982 New Mexico Geological Society Guidebook (p. 97-124), and Lozinsky (1988).

Stop 24 (S18)--Placitas Overlook

There is a discussion of this stop in Circular 163 (p. 158-162, 177-183). In particular, see discussion by V. C. Kelley (p. 159-162).

Stop 25 (J2)--Guadalupe Box

The Jemez Mountains are one of the two major volcanic centers in the northern Rio Grande rift. Volcanism in this area began more than 13 Ma ago and is anomalous in that intermediate to silicic magmas rather than basalts were involved. The major activity occurred in two phases and formed the Toledo and Valles calderas 1.5 Ma and 1.1 Ma ago. The Valles Caldera is the classic example of a resurgent caldera and two special sections of the Journal of Geophysical Research have been recently devoted to this area (February, 1986; June, 1988). This stop is on the southwest side of the Jemez Mountains just north of the Jemez Pueblo. Also see discussion of the southern Jemez Mountains by Bailey and Smith in Circular 163 (p. 184-187), Gardner et al. (1986), and Goff and Shevenell (1987).

Stop 26--New Mexico Museum of Natural History

This new museum contains many interesting exhibits on the geologic history of the region. The dinosaur displays are of particular interest.

Stop 27--Tijeras Canyon

Precambrian metamorphic terranes of Tijeras Canyon (from New Mexico Geological Society Guidebook, 33rd Field Conference, 1982, p. 4-6).

This stop is a turnoff to the site of an old Atomic Energy Commission radio tower and gives an excellent view of the Precambrian terranes as well as some of the Quaternary terrace gravels exposed along Tijeras Arroyo. The view to the southwest shows a panorama of Precambrian terranes: from left (southeast)--steep rim of southeast-tilted late Paleozoic strata, metavolcanic part of Tijeras greenstone, metasedimentary part of Tijeras greenstone with quartzite lenses, northeast-striking Tijeras fault, and meta-arkosic Cibola gneiss with resistant ridge of quartzite. Foliation and layering in both gneiss and greenstone strike generally northeast, with dips to the southeast in the greenstone and to the northwest in the gneiss.

The area is underlain by metasedimentary phyllite, quartzite, and metavolcanics of the Tijeras greenstone, and good exposures of highly deformed spotted phyllite may be seen about 300 m to the northeast along the gas pipeline service route. A drainage tunnel under I-40 about 170 m to the northeast from the pipeline route provides access to the Cibola gneiss terrane. A Tertiary, green lamprophyre dike with an altered chilled margin intrudes Cibola gneiss in a roadcut on I-40 to the northwest, and an excellent view of the Tijeras fault is visible in I-40 roadcuts about 0.7 km to the northeast.

Tijeras greenstone and Cibola gneiss are complexly deformed and metamorphosed Proterozoic terranes of contrasting lithology. Cibola gneiss is primarily meta-arkose with a prominent ridge of isoclinally folded quartzite, implying isoclinal folding has occurred throughout the terrane. Tijeras greenstone contain pelitic metasediments and quartzite in addition to the tholeiitic metavolcanics which dominate the terrane; isoclinal folding has also occurred in the greenstone. Syntectonic metamorphic grade in both terranes is low-pressure amphibolite facies, but the internal structural geometries in the two terranes are distinctly different.

The northeast-striking, subvertical Tijeras fault is buried beneath road fill in this area. Documented fault movement postdates folding and amphibolite-facies metamorphism in gneiss and greenstone, and the presence of numerous en-echelon granitic and pegmatitic dikes adjacent to the fault, in the greenstone southwest of Deadman's curve, suggests left-slip contemporaneous with granitic intrusion. The contrast in structure and lithology between the juxtaposed metamorphic terranes suggests considerable slip, probably at least several kilometers, has occurred on the Tijeras fault. There have been at least four episodes of movement along the fault, including left slip during the Precambrian, either dip-slip or strike-slip movement during middle and late Paleozoic time to form a north-facing scarp and related structure, late Cenozoic dip slip with subordinate left slip, and normal slippage (north-side down) during late Quaternary time. Also see Connolly and Kues (1987).

Stop 28--Madrid

Madrid and the Cerrillos coal field (Edward C. Beaumont, Consulting Geologist, Albuquerque, New Mexico, from New Mexico Geological Society Guidebook, 30th Field Conference, 1979, p. 269). Still fresh in the minds of oldtimers in Albuquerque and Santa Fe are the spectacular pageants and elaborate displays at Christmas in the once-booming mining town of Madrid. Now occupied by an ever-increasing number of relatively young people who are seeking principally a quieter and simpler life than can be found in the city, Madrid in its heyday was a major center of coal-mining activity.

A part of the Cerrillos coal field, the major Madrid Gulch development came rather late in the history of the greater area. Records indicate coal having been mined from the Madrid area to operate a gold mill at Dolores in 1835. There are also records (J. Huber, personal commun.) of coal having been mined from the "B" seam at the Government mine to supply military establishments at Santa Fe and Las Vegas during Civil War days.

It wasn't until what is now the Atchison, Topeka and Santa Fe Railway reached Cerillos (4.8 km north of Madrid) in 1880 that the coal field began to produce on an ongoing commercial basis. The earliest development apart from those mentioned above were in Waldo and Miller gulches which approximately parallel the north-trending Madrid Gulch, and are located about 1.6 km and 2.4 km west of Madrid Gulch, respectively. These earlier mining localities were closer to the main line of the railroad, and the coal is less involved with igneous intrusions that complicate the geology and hence the mining at Madrid. Also, the dip of the beds is several degrees less than in Madrid Gulch. The coal seams mined at Miller Gulch are lower in the Mesaverde sequence than those in Waldo and Madrid gulches. The Miller Gulch seam is reported to be about 1 m thick in an abandoned drift, and to thin to unminable thicknesses away from the Miller Gulch mine. The quality of the coal at Miller Gulch is such that it was in great demand for smithing and the manufacture of gas.

Initially, the Santa Fe spur was built from the Waldo switch across Galisteo Creek and up Waldo Gulch about 2.4 km to the old village of Waldo. The mines at Waldo operated by the railroad prospered for several years until the demand for anthracite became so great that the Santa Fe extended their spur up Madrid Gulch in 1889, and in 1893, they shifted their entire operation to Madrid. Generally, mining economics favored the Madrid anthracite operations, and the Miller and Waldo gulch operations were closed.

The Colorado Fuel and Iron Company of Pueblo, Colorado, leased the Santa Fe's holdings in 1896, and produced bituminous and anthracite coal from Madrid and a superior blending coal that was mixed with their Colorado coking coals. After ten years of difficult mining conditions resulting from steep dips, faults and dikes, Colorado Fuel and Iron relinquished their leases and the properties were leased to Mr. George Kaseman, a prominent Albuquerque businessman and banker. He operated the Albuquerque and Cerrillos Coal Company until his tragic death in 1938.

It was during this period that the Huber family, present owners of the property, came on the scene. Mr. Oscar Huber, who had been superintendent of the A&C operations, continued to operate the mines and eventually purchased the property in 1947. Economic events (e.g., conversion of the railroads to diesel power, and the advent of cheap natural gas and fuel oil) produced a gradual decline in Madrid production until the last mine closed in 1958.

The Cerillos coal field lies on the northwest flank of the Galisteo basin. The north-south striking Cretaceous strata dip eastward at an average of about 10°. At Madrid, however, the dip is somewhat greater, and in the mine entries, the dip ranges from 14° to 17°, decreasing basinward to less than 10°, as determined from elevations taken from mine maps. This stop is in the southern part of the region covered by the 1979 New Mexico Geological Society Guidebook (Day 2).

Stop 29 (S20)--La Bajada Escarpment

There is a discussion of this stop in Circular 163 (p. 165-167).

Stop 30--Santa Fe

Santa Fe is one of the oldest cities in

North America and one of the most interesting cities in the Southwest. It is a cultural center known for its shops, art galleries, restaurants, museums, and opera (Circular 163, p. 48-49, 167-169).

Stop 31 (E2)--Camel Rock

There is a discussion of this stop in Circular 163 (p. 201-203, p. 210-228).

Stop 32 (E4)--Santa Cruz Reservoir

There is a discussion of this stop in Circular 163 (p. 205, p. 210-228). Also see Menges (1987a) and Dethier et al. (1988).

Stop 33 (E7)--Truchas Rest Area

There is a discussion of this stop in Circular 163 (p. 206, p. 210-228).

Stop 34 (E10)--Southern Picuris Mountains

There is a discussion of this stop in Circular 163 (p. 207, p. 210-228). Also see 1979 and 1984 New Mexico Geological Society Guidebooks, and Bauer (1987) for detailed information on the Picuris Mountain-Harding Mine area.

Stop 35 (E11)--East of Dixon

There is a discussion of this stop in Circular 163 (p. 207, p. 210-228). Also see 1984 New Mexico Geological Society Guidebook (Day 1), and Stein Press (1981).

Stop 36 (N15)--Rio Grande Gorge at Pilar

There is a discussion of this stop in Circular 163 (p. 43-44). Also see Muehlberger (1979), Dungan et al. (1984), and Bauer (1987).

Stop 37 (N14)--Hondo Canyon Overlook

There is a discussion of this stop in Circular 163 (p. 45-46). Also see Muehlberger (1979), Dungan et al. (1984), Menges (1979a), and Wells et al. (1987).

Stop 38—Taos Pueblo

This pueblo is a record of man's inhabitation of the Rio Grande rift for at least 10,000 yrs. The Taos Pueblo is a classic example of the communal living style of the Pueblo Indians which inhabited the area when the Spaniards arrived in the 14th century (New Mexico Geological Society, 1984, p. 271-296).

Stop 39 (N13)--Rio Grande Gorge Bridge

There is a discussion of this stop in Circular 163 (p. 40-42). Also see Dungan et al. (1984), Menges (1987a, b), Wells et al. (1987), and Williams (1988).

Stop 40—Fort Garland

Fort Garland was established by the War Department in 1858 in what was then New Mexico Territory. The new fort replaced Fort Massachusetts, which was built in 1852 six miles north, but was vulnerable to hostile fire from high ground around the fort. The new fort was named for John Garland, army commander of the Department of New Mexico.

The purpose of the fort was to show an army presence in the area to deter Indian attacks on settlers in the San Luis Valley region. Men from Fort Garland also fought in campaigns in Arizona and Utah. Kit Carson, the famous scout and frontiersman, served as commandant of the fort. He was instrumental, because of his knowledge and the trust in him by the Indians, in keeping peaceful relations with the warlike Utes. At one time in the 1880s, 1,500 men were stationed at Fort Garland, but the garrison was generally less than 100. The fort was abandoned in 1883, as the threat of Indian raids decreased.

Stop 41—La Veta Pass/Spanish Peaks

The route to La Veta Pass is to the west and south of Blanca Peak and Mount Lindsey. The major part of these mountains are made up of Precambrian schists, gneisses, and intrusions. Also present are thrust slivers of Pennsylvanian rocks. One such sliver appears to be part of a lateral thrust ramp at California Gulch, north of Blanca Peak. A possible continuation of the California Gulch fault sliver of Pennsylvanian rocks underlies the upper part of Mount Lindsey. In both the California Gulch area and the Mount Lindsey area the Pennsylvanian rocks are sandwiched by thrusts so that Precambrian rocks lie both above and below (Tweto, 1979; Johnson et al., 1987).

As the route ascends La Veta Pass, it encounters progressively lower thrust plates. The road is in the lowest thrust plate, which puts Pennsylvanian rocks onto Cretaceous rocks at the pass.

The descent from La Veta Pass is near the Mount Mestas, a Tertiary micrograniitic intrusion. The route crosses the frontal thrust of the Sangre de Cristo thrust system on the south side of Mount Mestas going from Pennsylvanian rocks of the hanging wall onto Cretaceous in the foot wall. The Raton Basin into which the route descends was formed in Late Cretaceous and Early Tertiary time and may contain the most complete sedimentary record of the "Laramide" Orogeny of any foreland basin in the Rocky Mountains (Baltz, 1965). To the south is a view of the Spanish Peaks, two granite-granodiorite porphyry intrusions (Johnson, 1968; Knopf, 1936).

Stop 42—San Juan Sag, Hydrocarbon Potential

Scattered outcrops (Larsen and Cross, 1956; Sinclair, 1963; Tweto, 1970) and oil seeps in the fractured volcanic rocks of the San Juan field (Gries, 1985) indicate the presence of a Cretaceous sedimentary sequence concealed beneath the volcanic rocks. This interpretation is the basis for ongoing hydrocarbon exploration in the area. The San Juan Sag is separated from the San Juan Basin to the southwest, by the Archuleta arch.

It probably represents a relatively shallow intermontane basin formed during the Late Cretaceous-Early Tertiary Laramide Orogeny that was almost completely covered by Middle Tertiary volcanic rocks. Exploration drilling in the vicinity of Del Norte has revealed over 1,000 m of Cretaceous and Early Tertiary sedimentary rocks and hydrocarbon shows (Gries, 1985).

Stop 43—San Juan Volcanic Field

The San Juan volcanic field is made up largely of intermediate-composition lavas and volcaniclastic rocks, including silicic ash flow tuffs (Lipman et al., 1970, 1978). It represents the largest remnant of an extensive middle Tertiary volcanic complex (Steven, 1975). Activity began in Early Oligocene time with the eruption of voluminous intermediate composition lavas and breccias. These volcanic rocks built up a complex of stratovolcanoes in an area of more than 40,000 km^2 in which the average thickness of igneous rocks was approximately 1 km (Lipman et al., 1970).

Between approximately 30 m.y. and 26.5 m.y. ago, silicic ash flow tuff eruptions built a thick welded-tuff plateau overlying the earlier, intermediate composition volcanic complex. Some of the individual ash flow sheets were very large, covering up to 15,000 km^2 and having volumes estimated to have been 3,000 km^3. The tuffs were erupted from at least 18 calderas (Lipman et al., 1978).

The volcanic field today is marked by a large negative gravity anomaly that has been related to the presence of an epizonal batholith (Plouff and Pakiser, 1972; Berhendt and Bajwa, 1974). The earlier, intermediate volcanic and the later silicic ash flow tuffs may represent a fractionation of the magma body through time (Lipman et al., 1978).

Approximately 25 m.y. ago, the nature of the volcanism in the San Juan changed to widespread eruptions of basaltic lavas and local rhyolitic flows and tuffs (Lipman et al., 1978).

Stop 44—Great Sand Dunes

The Great Sand Dunes National Monument is located in the east-central part of the San Luis Valley. Eolian deposits blanket large areas of the eastern part of the basin, but it is the large dunes that have been set aside as a national monument.

The semi-arid climate of the rift in south-central Colorado (less than 20 cm/yr of rainfall in Alamosa, less than 25 cm/yr at the monument) supports sparse vegetation. Strong, consistent southwesterly winds blow across the floor of the San Luis Valley (Rio Grande fluvial fan), pick up sand from unconsolidated alluvial deposits in the basin, and carry it toward the Sangre de Cristo Mountains (Johnson, 1971). Strong turbulence, formed as the winds get close to the mountain front, reduces the capacity of the wind to move the sand, resulting in a deposit of sand. Transverse dunes form the largest (over 200 m high) eolian features. Many other dune types are present, depending on local wind and sand supply conditions. The dunes form in a reentrant in the mountain front. Medano Creek flowing westward out of the mountains serves to recycle some of the sand within the dune complex.

Stop 45—Window in the Sangre de Cristo Thrust System

The Sangre de Cristo Range from New Mexico to north of Crestone, Colorado, is an east-directed stack of thrust sheets (Lindsey et al., 1983; Johnson et al., 1987). These thrust sheets include Precambrian basement rocks in the structurally highest plates. Lower thrust plates consist of Pennsylvanian age arkosic and marine sedimentary rocks. Deformation took place during the Late Cretaceous-Early Tertiary Laramide Orogeny. The thrusting compressed a thick section that had been deposited in a narrow, linear, deep Pennsylvanian basin called the Central Colorado trough. This basin was compressed eastward against the buttress of the basement-cored Wet Mountain Uplift (Lindsey et al., 1983; Johnson et al., 1987).

The geometry of the thrust faults is well displayed in the Sangre de Cristo Mountains, but the amount of shortening is not well constrained. To the east and across the road from the Monument Visitor Center and the campground is a window through the Precambrian rocks in the hanging wall that exposed the Pennsylvanian rocks in the foot wall. The rocks in the foot wall are generally highly deformed and steeply dipping (Johnson et al., 1987).

This window exposes the foot wall beyond the crest of the range (Johnson et al., 1987). Depending on the interpretation of the structure in Huerfano Park and areas farther east of the range, up to 30 km of shortening is possible in the Precambrian rocks. Considerably more shortening is possible when thrusts in the Phanerozoic rocks are included.

Stop 46—Vallejo Formation (Eocene-?)

The Vallejo Formation (Upson, 1941; Huntley, 1979) is composed of red beds and conglomerates along the eastern side of the rift in Colorado. These rocks may represent the earliest deposits within the rift. They signal a change in the area from tectonic highland formed in the Late Cretaceous-Early Tertiary time to a basin receiving clastic, and in some cases coarse clastic, sediments. Unfortunately, the age of the Vallejo Formation is not well known. Upson (1941) interpreted the age of the unit to be Eocene to Oligocene, based on the fact that it is concordantly overlain by units containing volcanics of Oligocene age and on the fact that it rests unconformably on Precambrian rocks uplifted and denuded during the Late Cretaceous-Early Tertiary. Regardless of the age, it is not completely clear that these rocks are associated with faulting and rapid development of the Rio Grande Rift. Alternatively, they could represent the onlap of sedimentary rocks onto a degrading topographic highland. More work is required on these important rock units.

Stop 47 (N9)--Depth and Geometry of the Northern Rio Grande Rift in the San Luis Basin, South-Central Colorado

The deepest part of the northern Rio Grande Rift is just northwest of the Great Sand Dunes, on the eastern side of the San Luis Valley (Fig. 8). Approximately 150 km (95 mi) of high-quality, 20-fold CDP, seismic data (Fig. 9) indicate that the basin is filled with approximately 6.4 km (21,000 ft) of Tertiary sediments, mostly Oligocene or younger in age. These data, combined with published data, provide a basis for interpreting the structural geometry of the rift in this area (Kluth and Schaftenaar, 1989).

The internal structural geometry of this part of the rift (Fig. 10) is surprisingly uncomplicated for a feature so large and relatively young. Sympathetic and antithetic faulting is not widespread, or major, within the deep part of the basin. The internal geometry of the sedimentary packages, however, indicates a complicated movement history on the faults that are present within the rift.

The bounding fault zone between the rift and the Sangre de Cristo Mountains has approximately 9.2 km (30,000 ft) of vertical separation and may be very complicated. While Kluth and Schaftenaar (1989) were not able to resolve the bounding fault zone with their seismic data, these data provide constraints on the minimum angle of the fault zone at approximately 45°. Their gravity data (952 stations), combined with published data, indicate that the fault dips approximately 60° and is relatively steep to great depths (Circular 163, p. 26-28).

Stop 48 (N8)--Valley View Hot Springs and Northern San Luis Valley

The north end of the San Luis Valley narrows, and the structural geometry of the rift becomes more complex than in the south (Davis and Stoughton, 1979). In the vicinity of Villa Grove, Colorado, northwest striking faults of the Villa Grove Fault Zone cut Late Pleistocene alluvial material (McCalpin, 1983, 1987). The faults have formed a series of recognizable scarps, some of which are over 5 m high and most of which are over 1 m high. The young faulting indicates that deformation associated with rifting is continu-

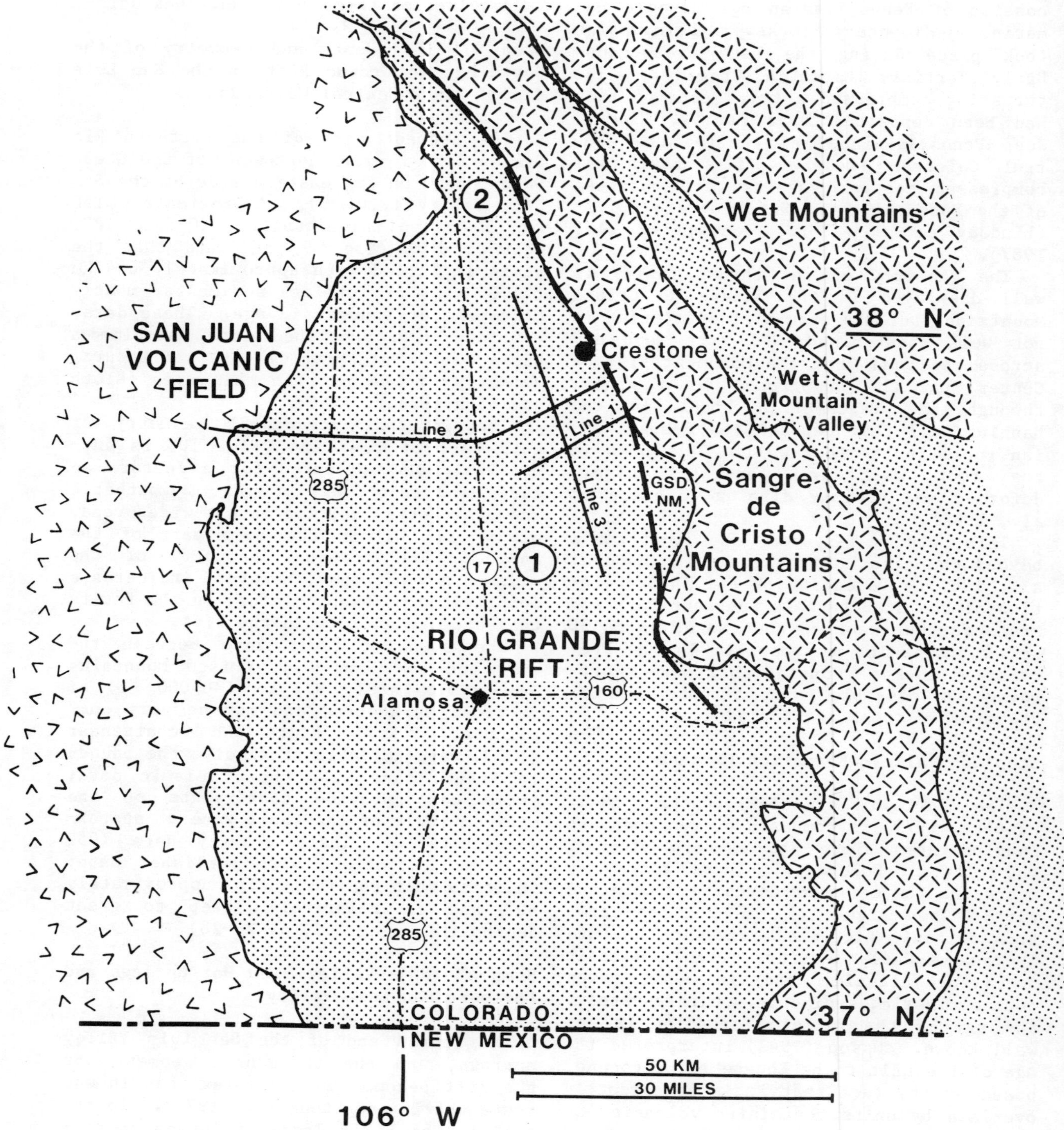

FIGURE 8 Index map of the San Luis Basin showing location of the seismic line (Line 2) which produced the record section showin in Figure 9. GSDNM--Great Sand Dunes National Monument. Numbers refer to previous seismic lines in the valley (see text). From Kluth and Schaftenaar (1989).

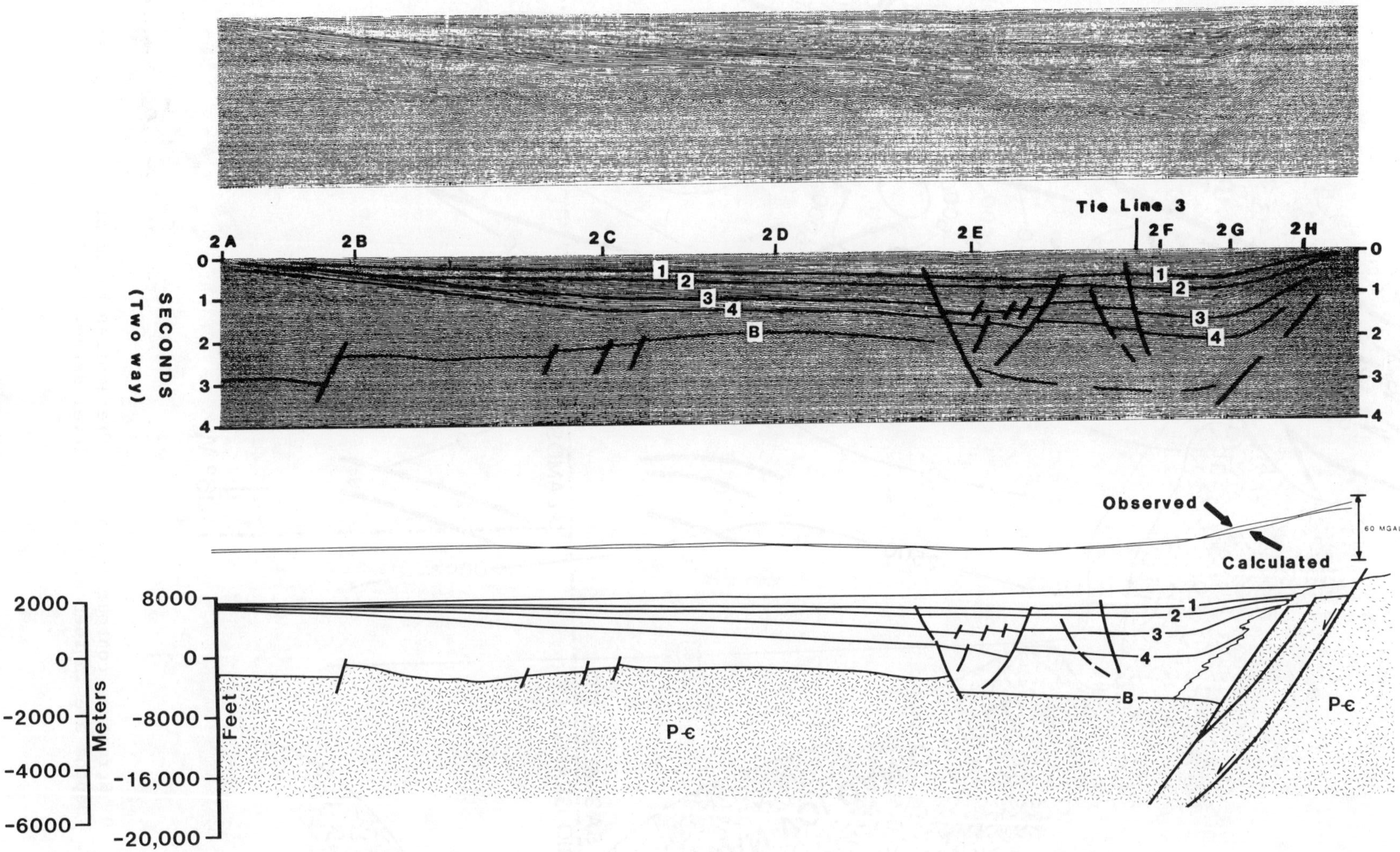

FIGURE 9 Seismic record section and interpretation of Line 2 (Fig. 8). Numbers indicate reflectors within the Cenozoic fill. From Kluth and Schaftenaar (1989).

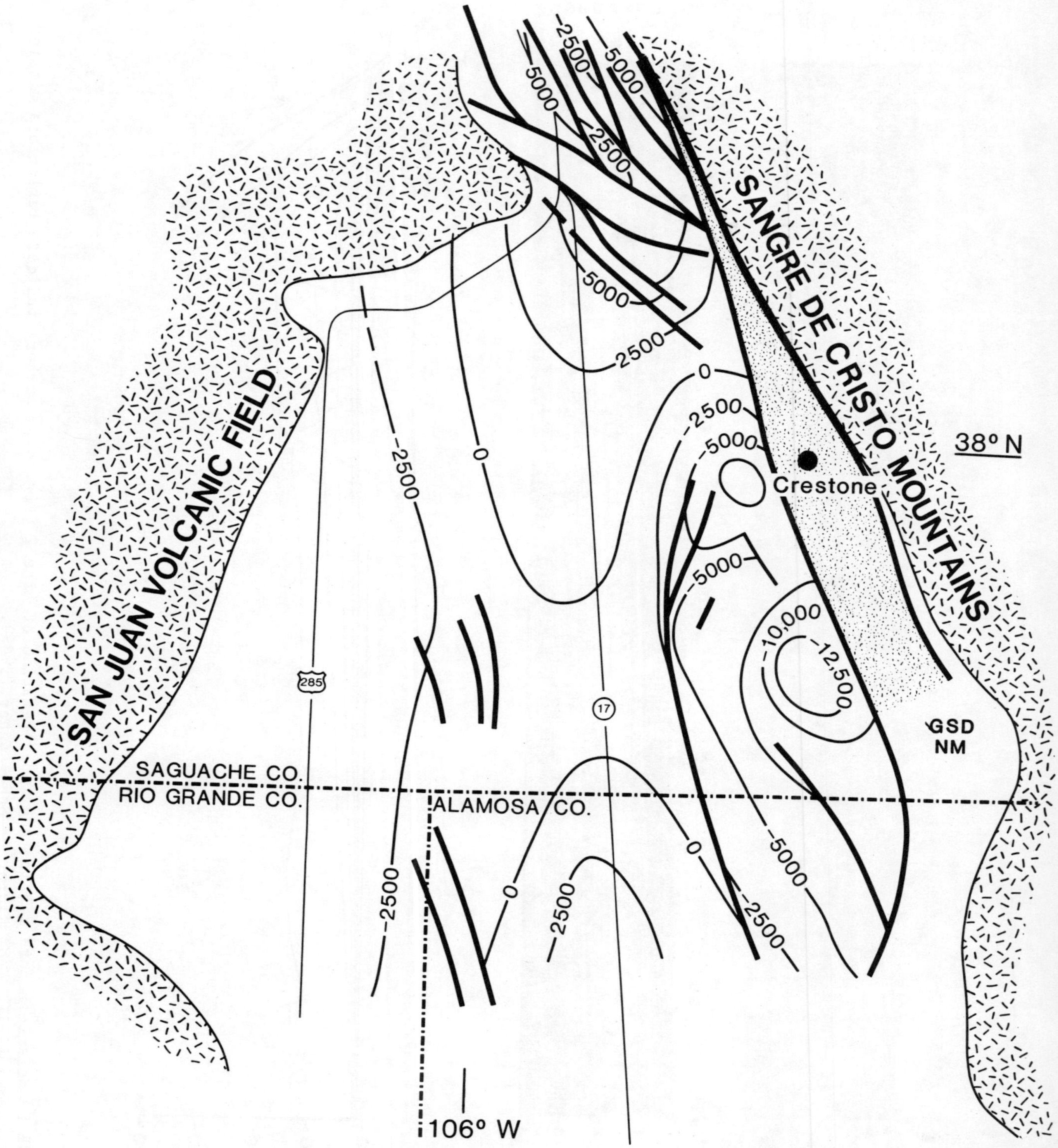

FIGURE 10 Structure contour map of the Precambrian basement in the northern San Luis Valley. Depths are in feet with a sea level datum.

ing. The scarps of the Villa Grove Fault Zone and those cutting alluvial fan deposits farther to the south, that also indicate young faulting, are of interest because seismicity in the Colorado part of the rift is much less than farther south in New Mexico. As yet, there is no accepted explanation for the lack of modern seismicity in this area of active tectonics and young faults (Coleman et al., 1985; McCalpin, 1983, 1987; Circular 163, p. 25-26).

Stop 49—Bonanza Mining District

The Bonanza mining district is within the northeastern part of the San Juan volcanic complex, and on the northwestern margin of the Rio Grande Rift. Mid-Tertiary (Oligocene) volcanic rocks rest within the area on Paleozoic sedimentary and Precambrian metamorphic rocks. The Precambrian rocks are exposed in only a few small areas in the district, but the Paleozoic sedimentary rocks, including lower Paleozoic sedimentary rocks not preserved in the Crestone area to the south, are exposed over a wide area. The Tertiary volcanic rocks are silicic to intermediate lavas, breccias, and tuffs common in the San Juan complex (Burbank, 1932; Tweto, 1979).

The area underwent folding and faulting, including thrusting during the Late Cretaceous (?) and Early Tertiary (?). The folds trend N-NW and, along Kerber Creek, the Paleozoic rocks are overridden by a thrust with Precambrian rocks on the hanging wall (Burbank, 1932; Tweto, 1979).

The volcanic rocks were erupted after extensive erosion, rest on the deformed older rocks, and produced an extensive plateau. This volcanic plateau was cut by faulting that generally dropped blocks toward a subsidence center located in the eastern part of the district (Burbank, 1932; Shannon et al., 1987). The fault blocks are rotated, perhaps related to an earlier doming that preceded subsidence (Burbank, 1932) or to movement on listric-shaped normal faults (see Burbank, 1932, Figure 6).

Stop 50—Poncha Pass

Poncha Pass is a topographically low area and at the northern end of the San Luis Basin part of the Rio Grande rift. The area is approximately 300 m higher than the San Luis Basin to the south, but is approximately 600 m above the Arkansas River Valley approximately 9 km to the north. Headward erosion by tributaries of the Arkansas River (Poncha and Little Cochetopa Creeks) is presently moving the surface drainage divide southward.

The area of Poncha Pass is a structurally complex, narrow trough containing Tertiary-age volcanics and sedimentary rocks. These Tertiary rocks are in depositional contact with Precambrian basement rocks on the east side of the trough, but are juxtaposed by faulting with the basement on the west side. The subsurface geometry of the faults is uncertain, but westward tilting of the Tertiary rocks with the trough suggest an east-side-down listric shape to the western bounding fault (Van Alstine, 1968).

The main part of the rift at Poncha Pass is a narrow trough that connects in an uncertain, but probably right en echelon, way to the Arkansas Valley to the north. Extension related to the rifting may be distributed over a wide area in the vicinity of Poncha Pass so that the physiographic expression of the rift is much more restricted in this en echelon transfer zone than to the north or south.

Stop 51—Arkansas River Canyon

The Arkansas River turns abruptly at the southern end of the Arkansas Valley from its southerly course to a southwesterly course. In doing so, it leaves the Rio Grande Rift and cuts through the Precambrian and Paleozoic rocks on the east flank of the Mosquito Range. This puzzling course is probably related to incisement of the river into the mid-Tertiary erosional surface and a complicated history of drainage capture.

The depositional contact of Lower Paleozoic sedimentary rocks on the Precambrian basement is exposed along the Arkansas River on the east flank of the Mosquito Range (Tweto, 1979). Here, the Lower Paleozoic rocks are overlain by coarser clastics of Pennsylvanian age. This relationship is different than to the south, in the Sangre de Cristo Range where the coarse Pennsylvanian rocks rest on the basement. In terms of Pennsylvanian paleogeography, the present Sangre de Cristo Range was on the flanks of a large basement cored uplift. Northward in the present-day Mosquito Range is part of the deep, narrow Central Colorado trough. Lower Paleozoic sediments were preserved in the Central Colorado trough, but not on the Pennsylvanian-age uplift to the south.

During the Late Cretaceous-Early Tertiary deformation, this trough was subjected to thrusting and folding.

Stop 52 (N6)--Arkansas Valley

The Arkansas Valley is the northernmost large graben related to the Rio Grande rift (Chapin, 1988). In this region, the rift has cut longitudinally through the east flank of a large anticline, the Sawatch Uplift, that formed during the Late Cretaceous-Early Tertiary Laramide Orogeny (Tweto, 1978, 1979). The Mosquito Range, composed of Precambrian basement and east-dipping Paleozoic rocks, to the east of the rift was the east flank of the large anticline. The Sawatch Range to the west of the rift is composed largely of Precambrian rocks in the core of the earlier anticline (Tweto, 1979).

Rifting began in the Arkansas Valley less than 29 Ma (Tweto, 1979). The graben is generally bounded by only one or two faults on its west side, but by many faults stepping up to the crest of the Mosquito Range on the east. Thickness of the graben fill (mostly Dry Union Formation) is approximately 1,000 m near Leadville (Tweto and Case, 1972; Tweto, 1979), 1,200 m near Buena Vista, and 1,500 m near Salida (Zohdy et al., 1971). Provenance of the Dry Union Formation and fault-cover relationships indicate that the east side of the graben was more active during the early history of the rifting and that the west side was more active later. The western bounding faults cut Quaternary deposits (Tweto, 1979).

The Sawatch Range on the west side of the rift has experienced rapid uplift and westward tilting during Late Tertiary time (Tweto, 1979). Several of the peaks in the range are over 4,500 m (14,000 ft) above sea level. Extensive hydrothermal activity has altered the crystalline rocks at Chalk Creek on the south flank of Mount Princeton (Tweto, 1979). The activity is continuing today (Circular 163, p. 22-23).

Stop 53 (N5)--West of Buena Vista

There is a discussion of this stop in Circular 163 (p. 21-22).

Stop 54 (N3)--Leadville and Climax

Leadville, located on the west flank of the Mosquito Range and on the east side of the Rio Grande Rift, is one of the richest mining districts in the U.S. Placer gold was discovered in the Leadville area in 1860 and lode ore in 1868. The cyclic history of the district is typical of most of the mining towns in Colorado and generally depended upon ore discoveries, transportation, prices, and development in smelting techniques. The value of the silver, zinc, lead, gold, and other metals is estimated at $525 million (Tweto, 1978b, c).

Mineralization in the Leadville District is generally caused by the alteration and reaction of Paleozoic carbonate rocks with mineralizing fluids. The primary ores are mixed sulphides in blanket alteration zones. The veins contain sulphides, but also silver, gold, and copper. Later Tertiary through at least Pleistocene faulting probably related to rifting has rotated blocks eastward and brought ore bodies to shallow levels within the district. These faults reactivate earlier faults formed during the late Cretaceous and Early Tertiary deformation that included the intrusive of porphyry sills and stocks. Later (probably post-Eocene [?] and pre-Miocene) hydrothermal activity emplaced the ores along contacts, fractures and faults (Tweto, 1978b, c; Emmons et al., 1927). Stratigraphy, structure, and intrusive features have all played an important role in mineralization (Circular 163, p. 18-19).

Stop 55 (N2)--Front Range--Climax

The Climax molybdenum district, approximately 20 kilometers NE of Leadville, is one of the richest deposits in the country. Over 60 million pounds of "moly" have been produced over the history of mining. The deposit was discovered within the highly iron-stained rocks of the area in 1879, but the ore mineral was not correctly identified until 1895. Until recently, production steadily increased to meet the growing demand for molybdenum.

Climax is not within the Rio Grande Rift proper. Normal faulting in the area, including that along the Mosquito Fault with approximately 3,000 m of west-side-down displacement, is probably related to nearby rifting (Wallace et al., 1978).

Mineralization is related to hydrothermal alteration in the vicinity of a middle Tertiary composite stock of rhyolite porphyry. Three major ore bodies are present related to three intrusive events. The ore is generally disseminated molybdenite, and includes huebnerite, cassiterite, pyrite, and monazite (Wallace et al.,

1978).

The northern terminus of grabens that define the Rio Grande rift (Chapin, 1971, 1988) is in the vicinity of Leadville. Large normal faults, such as the Mosquito Fault at Climax, are probably related to rifting. There are generally north-striking normal faults northward from Leadville to the Wyoming state line that may also relate to the Rio Grande Rift (Tweto, 1975; Circular 163, p. 18).

Stop 56 (N1)--Dillon Reservoir

There is a discussion of this stop in Circular 163 (p. 16-17).

Stop 57—Roadcut on I-70

This last stop is at a geologic point of interest on Interstate Highway 70 just west of Denver (see miles 11.1 to 12.2 in Circular 163, p. 14).

Acknowledgments

This field trip and guidebook are sponsored by the Rio Grande Rift Consortium which is an informal organization whose goals are to aid the dissemination of research results on the rift, encourage further research, and foster cooperation and communication. Two of the authors (G. Randy Keller and Kenneth H. Olsen) are involved in a National Science Foundation sponsored international study of continental rifts on a global scale. Some of the introductory material in this guidebook is a result of our efforts during this ongoing study, and this material includes input from Paul Morgan, Scott Baldridge, and Larry Braile.

REFERENCES

Aguilera, J. G., 1920, The Sonora earthquake of 1887, Bull. Seismol. Soc. Am., v. 10, 31-44.

Aiken, C. L. V., A. W. Laughlin, and F. A. West, 1978, Residual Bouguer gravity anomaly map of New Mexico. Los Alamos Nat. Lab., Pub. LA07466-MAP, 1:500,000.

Aldrich, M. J., Jr., C. E. Chapin, and A. W. Laughlin, 1986, Stress history and tectonic development of the Rio Grande Rift, New Mexico. J. Geophys. Res., v. 91, 6199-6211.

Ander, M. F., 1981, Geophysical study of the crust and upper mantle beneath the central Rio Grande Rift and adjacent Great Plains and Colorado Plateau. Los Alamos Nat. Lab. Rep. LA-8676-T, 218 p.

Ander, M. E., R. Goss and D. W. Strangway, 1984. A detailed magnetotellic/audiomagnetotelluric study of the Jemez Volcanic Zone, New Mexico: J. Geophys. Res., v. 89, 3335-3353.

Axelrod, D. I., and H. P. Bailey, 1976, Tertiary vegetation, climate, and altitude of the Rio Grande depression, New Mexico-Colorado. Paleobiology, v. 2, 235-254.

Baker, B. H., P. A. Mohr, and L. A. J. Williams, 1973, Geology of the Eastern Rift System of Africa. Geol. Soc. Am., Special Paper 136, 67 p.

Baldridge, W. S., 1979, Mafic and ultramafic inclusion suites from the Rio Grande rift (New Mexico) and their bearing on composition and thermal state of the lithosphere. Jour. Volcanology and Geothermal Research, v. 6, 319-351.

Baldridge, W. S., P. E. Damon, M. Shafiqullah, and R. J. Bridwell, 1980, Evolution of the central Rio Grande Rift, New Mexico: New potassium-argon ages. Earth Planet. Sci. Lett., v. 51, 309-321.

Baldridge, W. S., K. H. Olsen, and J. F. Callender, 1984, Rio Grande rift, problems and perspectives. New Mexico Geol. Soc., 35th Field Conf. Guidebook, 1-12.

Baldridge, W. S., F. V. Perry, and M. Shafiqullah, 1987, Late Cenozoic geology of the Lucero area, New Mexico. Geol. Soc. Amer. Bull., v. 99, 460-470.

Baltz, E. H., 1965, Stratigraphy and history of Raton Basin and notes on San Luis Basin, Colorado-New Mexico. Am. Assoc. Petrol. Geol. Bull., v. 49, 2041-2075.

Bauer, P. W., 1987, Precambrian geology of the Picuris Range, north-central New Mexico. New Mexico Bur. Mines Miner. Resour. Open-file Rept. 325, 280 p.

Behrendt, J. E., and L. Y. Bajwa, 1974, Bouguer gravity map of Colorado. U.S. Geol. Surv. Geophys., Invest. Map GP-895.

Birch, F. S., 1982, Gravity models of the Albuquerque basin, Rio Grande rift, New Mexico. Geophysics, v. 47, 1185-1197.

Black, B. A., 1984, Structural anomalies in the Espanola basin, New Mexico. Geol. Soc., 35th Field Conf. Guidebook, 59-62.

Blackwell, D. D., 1978, Heat flow and energy loss in the Western United States. In: Cenozoic Tectonic and Regional Geophysics of the Western Cordillera, R. B. Smith and G. P. Eaton (eds.), Geol. Soc. Am., Memoir, 152,

175-208.
Bridwell, R. J., 1976, Lithospheric thinning and the Late Cenozoic thermal and tectonic regime of the northern Rio Grande rift, Vermejo Park. New Mexico Geol. Soc., 27th Field Conf. Guidebook, 283-292.
Brown, L. D., P. A. Krumhansl, C. E. Chapin, A. R. Sanford, F. A. Cook, S. Kaufman, J. E. Oliver, and F. S. Schilt, 1979, COCORP seismic reflection studies of the Rio Grande rift. In: Rio Grande Rift: Tectonics and Magmatism, R. E. Riecker (ed.), Am. Geophys. Union, Spec. Pub., 169-184.
Brown, L. D., C. E. Chapin, A. R. Sanford, S. Kaufman, and J. Oliver, 1980, Deep structure of the Rio Grande rift from seismic reflection profiling. J. Geophys. Res., v. 85, 4773-4800.
Burbank, W. S., 1932, Geology and ore deposits of the Bonanza Mining District, Colorado: U.S. Geological Survey Professional Paper 189, 166 p.
Byerly, P., 1934a, The Texas earthquake of August 16, 1931. Bull. Seismol. Soc. Am., v. 24, 81-99.
Byerly, P., 1934b, The Texas earthquake of August 16, 1931. Bull. Seismol. Soc. Am., 303-325.
Case, J. E., and R. F. Sikora, 1984, Geologic interpretation of gravity and magnetic data in the Salida region, Colorado. U.S. Geol. Surv., Open-file Rpt. 84-372, 46 p.
Cash, D. J., and J. J. Wolff, 1984, Seismicity of the Rio Grande rift in northern New Mexico, 1973-1983. New Mexico Geol. Soc., 35th Field Conf. Guidebook, 25-28.
Chamberlin, R. M., 1980, Cenozoic stratigraphy and structure of the Socorro Peak volcanic center, central New Mexico: Unpublished Ph.D. dissertation, Colorado School of Mines, Golden, 495 p.; New Mexico Bureau of Mines and Mineral Resources, Open-file Report 118, 532 p., 3 pls.
Chamberlin, R. M., 1981, Cenozoic stratigraphy and structure of the Socorro Peak volcanic center, central New Mexico--a summary. New Mexico Geology, v. 3, no. 2, p. 22-24.
Chamberlin, R. M., 1983, Cenozoic domino-style crustal extension in the Lemitar Mountains, New Mexico--a summary. New Mexico Geological Society, Guidebook to 34th Field Conference, p. 111-118.
Chamberlin, R. M., and Osburn, G. R., 1986, Tectonic framework, character, and evolution of upper crustal extensional domains in the Socorro area of the Rio Grande rift, New Mexico. Arizona Geological Society Digest, vol. XVI, p. 464.
Chapin, C. E., 1971, The Rio Grande rift, Part I: Modifications and additions. New Mexico Geol. Soc., 22nd Field Conf. Guidebook, 191-202.
Chapin, C. E., 1988, Axial basins of the northern and central Rio Grande rift. In: Sedimentary Cover--North American Carton: U.S. (Basins of the Rocky Mountain Region), L. L. Sloss (ed.), Geol. Soc. Amer. (DNAG), the Geology of North America, v. D-2, 167-170.
Chapin, C. E., Chamberlin, R. M., Osburn, G. R., White, D. W., and Sanford, A. R., 1978, Exploration framework of the Socorro geothermal area, New Mexico: In: Field Guide to Selected Cauldrons and Mining Districts of the Datil-Mogollon Volcanic Field, New Mexico C. E. Chapin and W. E. Elston (eds.), New Mexico Geological Society, Special Publication No. 7, p. 114-129.
Clarkson, G., and M. Reiter, 1984, Analysis of terrestrial heat flow profiles across the Rio Grande rift and southern Rocky Mountains in northern New Mexico. New Mexico Geol. Soc., 35th Field Conf. Guidebook, v. 35, 39-44.
Coleman, S. M., J. P. McCalpin, D. A. Ostenaa, and R. M. Kirkham, 1985, Map showing upper Cenozoic rocks and deposits and Quaternary faults, Rio Grande rift, south-central Colorado: U.S. Geol. Surv. Misc. Invest. Map I-1594, scale 1:125,000.
Connolly, J. R., and Kues, B. S., Precambrian-Upper Paleozoic geology along I-40 east of Albuquerque, New Mexico. Rocky Mtn. Sec. Geol. Soc. Amer., Centennial Field Guide, v. 2, 437-440.
Cook, F. A., E. R. Decker, and S. B. Smithson, 1978, Preliminary transient heat flow model of the Rio Grande rift in southern New Mexico. Earth Planet. Sci. Lett., v. 40, 316-326.
Cook, F. A., D. B. McCuller, E. R. Decker, and S. B. Smithson, 1979, Crustal structure and evolution of the southern Rio Grande rift. In: Rio Grande Rift: Tectonics and Magmatism, R. E. Riecker (ed.), Am. Geophys. Spec. Pub., 195-208.
Cordell, L., 1976, Aeromagnetic and gravity studies of the Rio Grande graben in New Mexico between Belen and Pilar. New Mexico Geol. Soc., Spec. Publ. 6, 62-70.
Cordell, L., 1978. Regional geophysical setting of the Rio Grande rift. Geol. Soc. Am. Bull., v. 89, 1073-1090.

Cordell, L., 1982, Extension in the Rio Grande rift. J. Geophys. Res., v. 87, 8561-8569.

Cordell, L., 1984, Composite residual total intensity aeromagnetic map of New Mexico. National Oceanographic and Atmospheric Admin., National Geophys. Data Center Map, 1:500,000.

Cordell, L., G. R. Keller, and T. G. Hildenbrand, 1982, Bouguer gravity map of the Rio Grande rift. U.S. Geol. Surv., Map GP-949, 1:1,000,000.

Cordell, L., and Keller, G. R., 1984, Regional structural trends inferred from gravity and aeromagnetic data in the New Mexico-Colorado border region, New Mexico Geol. Soc. Guidebook, 35th Field Conf., 21-23.

Daggett, P. H., 1982, An integrated geophysical study of the crustal structure of the southern Rio Grande rift. Ph.D. Thesis, New Mexico State Univ., Las Cruces, 197 p.

Daggett, P. H., G. R. Keller, P. Morgan, and C. L. Wen, 1986, Structure of the southern Rio Grande rift from gravity interpretation. J. Geophys. Res., v. 91, 6157-6167.

Davis, P. M., E. C. Parker, Evans, J. R., H. M. Iyer, and K. H. Olsen, 1984, Teleseismic deep sounding of the velocity structure beneath the Rio Grande rift. New Mexico Geol. Soc., 35th Field Conf. Guidebook, 29-38.

Davis, T. L., and D. Stoughton, 1979, Interpretation of seismic reflection data from the northern San Luis Valley, south-central Colorado. In: Rio Grande Rift: Tectonics and Magmatism, R. E. Reicker (ed.), Am. Geophys. Union, Spec. Publ., 185-194.

DeAngelo, M. V., and Keller, G. R., 1988, Geophysical anomalies in southwestern New Mexico. New Mexico Geol. Soc. Guidebook, 39th Field Conf., 71-75.

Decker, E. R., and S. B. Smithson, 1975, Heat flow and gravity interpretation across the Rio Grande rift in southern New Mexico and west Texas. J. Geophys. Res., v. 80, 2542-2552.

Decker, E. R., G. J. Bucher, K. L. Buelow, and H. P. Heasler, 1984, Preliminary interpretation of heat flow and radioactivity in the Rio Grande rift zone in central and northern Colorado. New Mexico Geol. Soc. 35th Field Conf. Guidebook, 45-50.

Dethier, D. P., C. D. Harrington, and M. J. Aldrich, 1988, Late Cenozoic rates of erosion in the western Espanola basin, New Mexico: Evidence from geologic dating of erosion surfaces. Geol. Soc. Amer. Bull., v. 100, 928-937.

DeVoogd, B., L. D. Brown, and C. Merey, 1986, Nature of the eastern boundary of the Rio Grande rift from COCORP surveys in the Albuquerque basin, New Mexico. J. Geophys. Res., v. 91, 6305-6320.

Doser, D. I., 1987, The 16 August 1931 Valentine, Texas, earthquake: Evidence for normal faulting in west Texas. Bull. Seismol. Soc. Am., v. 77, 2005-2017.

Dumas, D. B., 1981, Seismicity of west Texas. Ph.D. Dissertation, Univ. of Texas at Dallas, Dallas, TX, 94 p.

Dumas, D. B., H. J. Dorman, and G. V. Latham, 1980, A reevaluation of the August 16, 1931 Texas earthquake. Bull. Seismol. Soc. Am., v. 70, 1171-1180.

Dungan, M. A., W. R. Muehlberger, L. Leininger, C. Peterson, N. J. McMillan, G. Gunn, M. Lindstrom, and L. Haskin, 1984, Volcanic and sedimentary stratigraphy of the Rio Grande gorge and the late Cenozoic geologic evolution of the southern San Luis Valley. New Mex. Geol. Soc., 35th Field Conference Guidebook, 157-170.

Eaton, G. P., 1986, A tectonic redefinition of the southern Rocky Mountains. Tectonophysics, v. 132, 163-193.

Edwards, C. L., M. Reiter, C. Shearer, and W. Young, 1978, Terrestrial heat flow and crustal radioactivity in northeastern New Mexico and southeastern Colorado. Geol. Soc. Am. Bull., v. 89, 1341-1350.

Emmons, S. F., Irving, J. D., and Laughlin, G. F., 1927, Geology and ore deposits of the Leadville mining district, Colorado. U.S. Geological Survey Professional Paper 148, 368 p.

Figures, S. H., 1987, Structural geology and geophysics of the pipline complex, northern Franklin Mountains, El Paso, Texas. Ph.D. Dissertation, Univ. Texas at El Paso, El Paso, TX, 279 p.

Foley, L. L., R. C. La Forge, and L. A. Piety, 1988, Seismotectonic study for Elephant Butte and Caballo Dams, Rio Grande Project, New Mexico. U.S. Bur. Reclamation, Div. of Geotechnical Engineering and Geology, Seismotectonic Rept. 88-9, 60 p.

Gardner, J. N., F. Goff, S. Garcia, and R. C. Hagan, 1986, Stratigraphic relations and lithologic variations in the Jemez volcanic field, New Mexico. Jour. Geophys. Res., v. 91, 1763-1778.

Gile, L. H., J. W. Hawley, and R. B. Grossman, 1981, Soils and geomorphology in the Basin and Range area of southern New Mexico. New Mexico Bur. Mines Miner. Resour. Memoir 39, 222 p.

Gilmer, A. L., R. A. Mauldin, and G. R. Keller, 1986, A seismic reflection study of the Jornada del Muerto and Palomas basins. New Mexico Geol. Soc., 37th Field Conf. Guidebook, 131-134.

Goff, F., and Shevenell, L., 1987, Travertine deposits of Soda Dam, New Mexico and evolution of the Valles caldera hydrothermal system. Geol. Soc. Amer. Bull., v. 99, 292-302.

Gough, D. I., 1974, Electrical conductivity under western North America in relation to heat flow, seismology, and structures. J. Geomagn. Geoelctr., v. 26, 105-123.

Gries, R. R., 1985, San Juan Sag: Cretaceous rocks in a volcanic covered basin, south central Colorado. Mountain Geologist, v. 22, p. 167-179.

Hawley, J. W. (comp.), 1978, Guidebook to the Rio Grande rift in New Mexico and Colorado, New Mex. Bur. Mines Miner. Resour. Circular 163, 241 p.

Hawley, J. W., 1984, Hydrogeologic cross sections of the Mesilla Bolson, New Mexico and Texas. New Mexico Bur. Mines Miner. Resour. Open-file Rept. 190, 10 p.

Henry, C. D., and J. G. Price, 1985, Summary of the tectonic development of Trans-Pecos Texas. Bur. Econ. Geol., Austin, Texas, Misc. Map No. 36.

Henry, C. D., and J. G. Price, 1986, Early basin and range development in Trans-Pecos Texas and adjacent Chihuahua: Magmatism and orientation, timing, and style of extension. Jour. Geophys. Res., v. 91, 6213-6224.

Hermance, J. F., and J. Pedersen, 1980, Deep structure of the Rio Grande rift: A magnetotelluric interpretation. J. Geophys. Res., v. 85, 3899-3912.

Huntley, David, 1979, Cenozoic faulting and sedimentation in northern San Luis Valley, Colorado: Summary. Geol. Soc. of Amer. Bull., Part I, v. 90, p. 8-10.

Jaksha, L. H., 1982, Reconnaissance seismic refraction-reflection surveys in southwestern New Mexico. Geol. Soc. Amer. Bull., v. 93, 1030-1037.

Jaksha, L. H. and Sanford, A. R., 1986, Earthquakes near Albuquerque, New Mexico, 1976-1981. Jour. Geophys. Res., v. 91, 6293-6303.

Jiracek, G. R., Ander, M. E., and Holocombe, H. T., 1979, Magnetotelluric soundings of crustal conductive zones in major continental rifts. In: Rio Grande Rift: Tectonics and Magmatism, R. E. Riecker (ed.), Amer. Geophys. Un., Spec. Publ., 209-222.

Jiracek, G. R., E. Gustafson, and P. S. Mitchell, 1983, Magnetotelluric results opposing magma origin of crustal conductors in the Rio Grande rift. Tectonophysics, v. 94, 299-326.

Jiracek, G. R., W. L. Rodi, and L. L. Vanyan, 1987, Implications of magnetotelluric modeling for the deep crustal environment in the Rio Grande rift. Phys. Earth Plan. Interiors, v. 45, 179-192.

Johnson, R. B., 1968, Geology of the igneous rocks of the Spanish Peaks region, Colorado. U.S. Geol. Surv. Prof. Paper 594-G, 47 p.

Johnson, R. B., 1971, The Great Sand Dunes of southern Colorado. New Mexico Geol. Soc. 22nd Field Conference Guidebook, 123-128.

Johnson, B. R., Lindsey, D. A., Bruce, R. M., Soullier, S. J., 1987, Reconnaissance geologic map of the Sangre de Cristo Wilderness Study Area, south-central Colorado. U.S. Geological Survey Map MF-1635-B.

Keller, G. R., L. W. Braile, and P. Morgan, 1979, Crustal structure, geophysical models and contemporary tectonism of the Colorado Plateau. Tectonophysics, v. 61, 131-147.

Keller, G. R., and L. Cordell, 1984, Bouguer gravity anomaly map of New Mexico. National Oceanic and Atmospheric Admin. National Geophys. Data Center Map, 1:500,000

Keller, G. R., L. Cordell, G. H. Davis, W. J. Peeples, and G. White, 1984, A geophysical study of the San Luis Basin. New Mexico Geol. Soc. 35th Field Conf. Guidebook, 51-57.

Keller, G. R., R. A. Smith, W. J. Hinze, and C. L. V. Aiken, 1985, A regional gravity and magnetic study of West Texas. Soc. Explor. Geophys., Spec. Pub., Utility of Regional Gravity and Magnetic Anomaly Maps, 198-212.

Keller, G. R., W. R. Seager, and S. Thompson, 1986, A seismic reflection study of part of the southern Jornada del Muerto. New Mexico Geol. Soc., 37th Field Conf. Guidebook, 139-142.

Keller, G. R., P. Morgan, and W. R. Seager, 1988, Crustal structure, gravity anomalies, and heat flow in the southern Rio Grande rift and their relationship to extensional tectonics. Tectonophysics, in press.

Kelley, S. A., and I. J. Duncan, 1984,

Tectonic history of the northern Rio Grande rift derived from apatite fission-track geochronology. New Mexico Geol. Soc., 35th Field Conf. Guidebook, 67-73.

Kelley, S. A., and I. J. Duncan, 1986, Late Cretaceous to middle Tertiary tectonic history of the northern Rio Grande rift. J. Geophys. Res., v. 91, 6246-6262.

Kluth, C. F., and C. H. Schaftenaar, 1989, Depth and geometry of the northern Rio Grande rift in the San Luis basin, south-central Colorado. Submitted to Geol. Soc. Am. Bull.

Knopf, Adolf, 1936, Igneous geology of the Spanish Peaks region, Colorado: Geol. Soc. of Amer. Bull., v. 47, 1727-1784.

Larsen, E. S., Jr., and Cross, Whitman, 1956, Geology and petrology of the San Juan region, southwestern Colorado: U.S. Geological Survey Professional Paper 258, 303 p.

Larsen, S. C., and R. E. Reilinger, 1983, Recent measurements of crustal deformation related to the Socorro magma body. New Mexico Geol. Soc., 34th Field Conf. Guidebook, v. 34, 119-121.

Larsen, S. C., R. E. Reilinger, and L. Brown, 1986, Evidence of ongoing crustal deformation related to magmatic activity near Socorro, New Mexico. J. Geophys. Res., v. 91, 6283-6292.

LeMone, D. V., and W. C. Cornell, 1988, Stratigraphy field guide, Franklin Mountains, Texas and New Mexico. El Paso Geol. Soc., 43 p.

Lindsey, D. A., Johnson, B. R., and Andriessen, P. A. M., 1983, Laramide and Neogene structure of the northern Sangre de Cristo Range, south-central Colorado. In: Rocky Mountain Foreland Basins and Uplifts, Rocky Mountain Association of Geologists, J. D. Lowell (ed.), 219-228.

Lipman, P. W., 1969, Alkalic and tholeiitic basalt volcanism related to the Rio Grande depression, southern Colorado and northern New Mexico. Geol. Soc. Am. Bull., v. 80, 1343-1354.

Lipman, P. W., Doe, B. R., Hedge, C. E., and Steven, T. A., 1978, Petrologic evolution of the San Juan volcanic field, southwestern Colorado: Pb and Sr isotope evidence. Geol. Soc. Am. Bull., v. 89, 59-82.

Lipman, P. W., and H. H. Mehnert, 1975, Late Cenozoic basaltic volcanism and the development of the Rio Grande depression in the southern Rocky Mountains. In: Cenozoic History of the Southern Rocky Mountains. Geol. Soc. Am. Memoir 144, 119-154.

Lipman, P. W., Steven, T. A., and Mehnert, H. H., 1970, Volcanic history of the San Juan Mountains, Colorado, as indicated by potassium-argon dating. Geol. Soc. Am. Bull., v. 81, 2329-2352.

Long, C. L., 1985, Regional audiomagnetotelluric study of the Questa caldera, New Mexico. J. Geophys. Res., v. 90, 11270-11274.

Lozinsky, R. P., 1986, Geology and late Cenozoic history of the Elephant Butte area, Sierra County, New Mexico: New Mexico Bureau of Mines and Mineral Resources, Circular 187, 40 p.

Lozinsky, R. P., 1987, Cross section across the Jornada del Muerto, Engle, and northern Palomas basins, south-central New Mexico. New Mexico Geology, p. 55-57.

Lozinsky, R. P., 1988, Stratigraphy, sedimentology, and sand petrology of the Santa Fe Group and pre-Santa Fe Tertiary deposits in the Albuquerque basin, central New Mexico (Ph.D. Dissertation): Socorro, New Mexico Institute of Mining and Technology, 298 p.

Lozinsky, R. P., and Hawley, J. W., 1986, Upper Cenozoic Palomas Formation of south-central New Mexico. New Mexico Geological Society, 37th Field Conference, Guidebook, p. 239-247.

Lozinsky, R. P., Kelly, J. C., Osburn, G. R., and Kudo, A. M., 1986, Boat log for the southern half of Elephant Butte reservoir. New Mexico Geological Society, 37th Field Conference, Guidebook, p. 57-65.

Machette, M. N., 1986, History of Quaternary offset and paleoseismicity along the La Jencia fault, central Rio Grande rift, New Mexico. Bull. Seism. Soc. Am., v. 76, 259-272.

Machette, M. N., 1987a. Preliminary assessment of paleoseismicity at White Sands Missile Range, southern New Mexico: Evidence for recency of faulting, fault segmentation, and repeat intervals for major earthquakes in the region. U.S. Geol. Surv., Open-file Rpt. 87-444, 46 p.

Machette, M. N., 1987b, Preliminary assessment of Quaternary faulting near Truth or Consequences, New Mexico. U.S. Geological Survey Open-file Report 87-652, 40 p.

Mattick, R. E., 196, A seismic and gravity profile across the Hueco bolson. Texas U.S. Geol. Surv. Prof. Paper 575-D, 85-91.

McCalpin, J., 1983, Quaternary geology and

neotectonics of the west flank of the northern Sangre de Cristo Mountains, south-central Colorado, Colorado School of Mines Quart., v. 77, 97 p.

McCalpin, J. P., 1987, Recurrent Quaternary faulting at Major Creek, Colorado. Rocky Mtn. Sec. Geol. Soc.Amer. Centennial Field Guide, v. 2, 353-356.

McGrath, D. B., and Hawley, J. W., 1987, Geomorphic evolution and soil-geomorphic relationships in the Socorro area, central New Mexico. In: Guidebook to the Socorro Area, V. T. McLemore and M. R. Bowie (eds.), New Mexico Bur. Mines Miner. Resour. Guidebook, 55-67).

McLemore, V. T., and Bowie, M. R. (eds.), 1987, Guidebook to the Socorro area, New Mexico. New Mexico Bur. Mines Miner. Resour. Guidebook, 75 p.

Menges, C. M. (ed.), 1987a, Quaternary tectonics, landform evolution, soil chronologies, and glacial deposits--northern Rio Grande rift of New Mexico. Friends of the Pleistocene-Rocky Mountain Cell. 1987 Field Trip Guidebook (Univ. New Mexico Geol. Dept.), 213 p.

Menges, C. M., 1987b, Temporal and spatial segmentation of Plio-Quaternary fault rupture along the base of the western Sangre de Cristo Mountain front. In: Quaternary Tectonics, Landform Evolution, Soil Chronologies, and Glacial Deposits--Northern New Mexico. Friends of the Pleistocene Rocky Mountain Cell, 1987 Field Trip Guidebook, 71-93.

Morgan, P., and M. P. Golombek, 1984, Factors controlling the phases and styles of extension in the northern Rio Grande rift. New Mexico Geol. Soc., 35th Field Conf. Guidebook, 13-20.

Morgan, P., and C. A. Swanberg, 1985, On the Cenozoic uplift and tectonic stability of the Colorado Plateau. Geodynamics, v. 3, 39-63.

Morgan, P., W. R. Seager, and M. P. Golombek, 1986, Cenozoic thermal, mechanical, and tectonic evolution of the Rio Grande rift. J. Geophys. Res., v, 91, 6263-6276.

Muehlberger, W. R., 1979, The Embudo fault between Pilar and Arroyo Hondo, New Mexico; an active intracontinental transform fault. New Mexico Geol. Soc., 30th Field Conference Guidebook, 77-82.

Muehlberger, W. R., R. C. Belcher, and L. K. Goetz, 1978, Quaternary faulting in Trans-Pecos Texas. Geology, v. 6, 337-340.

Ni, J. F., R. E. Reilinger, and L. D. Brown, 1981, Vertical crustal movements in the vicinity of the 1931 Valentine, Texas, earthquake. Bull. Seismol. Soc. Am., v. 71, 857-864.

Olsen, K. H., G. R. Keller, and J. N. Stewart, 1979, Crustal structure along the Rio Grande rift from seismic refraction profiles. In: Rio Grande Rift: Tectonics and Magmatism, R. E. Riecker (ed.), Am. Geophys. Union, Spec. Publ., 127-143.

Olsen, J. H., W. S. Baldridge, and J. F. Callender, 1987, Rio Grande rift: An overview. Tectonophysics, v. 143, 119-139.

Osburn, G. R., and Chapin, C. E., 1983, Nomenclature for Cenozoic rocks of northeast Mogollon-Datil volcanic field, New Mexico Bureau of Mines and Mineral Resources, Stratigraphic Chart 1.

Ouchi, S., 1983, Effects of uplift on the Rio Grande over the Socorro magma body, New Mexico. New Mexico Geol. Soc., 34th Field Conf. Guidebook, 54-56.

Padovani, E. R., and Carter, J. L., 1977, Aspects of the deep crustal evolution beneath south-central New Mexico. Am. Geophys. Un. Monograph 20, 19-25.

Parker, E. C., P. M. Davis, J. R. Evans, H. M. Iyer, and K. H. Olsen, 1984, Upwarp of anomalous asthenosphere beneath the Rio Grande rift. Nature, v. 312, 354-356.

Perry, F. V., W. S. Baldridge, and D. J. DePaolo, 1987, Role of asthenosphere and lithosphere in the genesis of Late Cenozoic basaltic rocks from the Rio Grande rift and adjacent regions of southwestern United States. Jour. Geophys. Res., v. 92, 9193, 9213.

Perry, F. V., W. S. Baldridge, and D. J. DePaolo, 1988, Chemical and isotopic evidence for lithospheric thinning beneath the Rio Grande rift. Nature, v. 332, 432-434.

Plouff, D., and L. C. Pakiser, 1972, Gravity study of the San Juan Mountains, Colorado. U.S. Geol. Surv. Prof. Paper 800-B, B183-B190.

Porath, H., and D. I. Gough, 1971, Mantle conductive structures in the western United States from magnetometer array studies. Geophys. J. Roy. Astr. Soc., v. 22, 261-275.

Pray, L. C., 1961, Geology of the Sacramento Mountains escarpment, New Mexico. New Mexico Bur. Mines Miner. Resour. Bull. 35, 144 p.

Prodehl, C., and L. C. Pakiser, 1980, Crustal structures of the southern Rocky Mountains from seismic measurements. Geol. Soc. Am. Bull., Pt. 1, v. 91, 147-155.

Ramberg, I. B., F. A. Cook, and S. B. Smithson, 1978, Structure of the Rio Grande rift in southern New Mexico and west Texas based on gravity interpretation. Geol. Soc. Am. Bull., v. 89, 107-123.

Reid, H. F., 1911, Remarkable earthquakes in central New Mexico in 1906 and 1907. Bull. Seismol. Soc. Am., v. 1, 10-16.

Reilinger, R. E., and J. E. Oliver, 1976, Modern uplift associated with a proposed magma body in the vicinity of Socorro, New Mexico. Geology, v. 4, 583-586.

Reilinger, R. E., J. E. Oliver, L. D. Brown, A. R. Sanford, and E. I. Balazs, 1980a, New measurements of crustal doming over the Socorro magma body. Geology, v. 8, 291-295.

Reilinger, R. E., L. Brown, and D. Powers, 1980b, New evidence for tectonic uplift in the Diablo Plateau region, west Texas. Geophys. Res. Lett., v. 7, 181-184.

Reiter, M., C. L. Edwards, H. Hartman, and C. Weidman, 1975, Terrestrial heat flow along the Rio Grande rift, New Mexico and southern Colorado. Geol. Soc. Am. Bull., v. 86, 811-818.

Reiter, M., A. J. Mansure, and C. Shearer, 1979, Geothermal characteristics of the Rio Grande rift within the southern Rocky Mountains complex 1979. In: Rio Grande Rift: Tectonics and Magmatism, R. E. Riecker (ed.), Am. Geophys. Union, Spec. Pub., 253-267.

Reiter, M., and J. C. Tovar, 1982, Estimates of terrestrial heat flow in northern Chihuahua, Mexico, based on petroleum bottom hole temperatures. Geol. Soc. Am. Bull., v. 93, 613-624.

Reiter, M., R. E. Eggleston, B. R. Broadwell, and J. Minier, 1986, Terrestrial heat flow estimates from deep petroleum tests along the Rio Grande rift in central and southern New Mexico. J. Geophys. Res., v. 91, 6225-6245.

Riecker, R. E. (ed.), 1979, Rio Grande rift: Tectonics and magmatism. Am. Geophys. Union, Spec. Pub., 438 p.

Rinehart, E. J., A. R. Sanford, and R. M. Ward, 1979, Geographic extent and shape of an extensive magma body at midcrustal depths in the Rio Grande rift near Socorro, New Mexico. In: Rio Grande Rift: Tectonics and Magmatism, R. E. Riecker (ed.), Am. Geophys. Union, Spec. Publ., 237-251.

Rodriguez, C., M. Martinez, R. Fernandez, J. M. Romo, and G. Jiracek, 1987, A 2-D magnetotelluric interpretation and its comparison with other geophysical methods for an east portion of the Espanola basin in the Rio Grande rift. Intl. Un. Geodesy Geophys., Abstr., 2, 443.

Sanford, A. R., A. J. Budding, J. P. Hoffman, O. S. Alptekin, C. A. Rush, and T. R. Toppozada, 1972, Seismicity of the Rio Grande rift in New Mexico. New Mexico Bur. Mines and Min. Res. Circ. 120.

Sanford, A. R., O. Alptekin, and T. R. Toppozada, 1973, Use of reflection phases on microearthquakes seismograms to map an unusual discontinuity beneath the Rio Grande rift. Bull. Seism. Soc. Am., v. 63, 2021-2034.

Sanford, A. .R., R. P. Mott, P. J. Schuleski, E. J. Rinehart, F. J. Caravella, R. M. Ward, and T. C. Wallace, 1977, Geophysical evidence for a magma body in the crust in the vicinity of Socorro, New Mexico. In: The Earth's Crust: Its Nature and Physical Properties, J. G. Heacoc (ed.), Am. Geophys. Union, Geophys. Monogr. 20, 385-403.

Sanford, A. R., R. P. Mott, Jr., P. J. Shuleski, E. J. Rinehart, F. J. Caravella, R. M. Ward, and T. C. Wallace, 1979, Geophysical evidence for a magma body in the crust in the vicinity of Socorro, New Mexico. In: Rio Grande Rift: Tectonics and Magmatism, R. E. Riecker (ed.), Am. Geophys. Union, Spec. Pub., 385-403.

Sass, J. H., D. D. Blackwell, D. S. Chapman, J. K. Costrain, E. R. Decker, L. A. Lawver, and C. A. Swanberg, 1981, Heat flow from the crust of the United States. In: Physical Properties of Rocks and Minerals, Y. S. Touloukian, W. R. Judd, and R. F. Roy (eds.), McGraw-Hill, New York, 503-548.

Savage, J. C., M. Lisowski, W. H. Prescott, and A. R. Sanford, 1980, Geodetic measurement of horizontal deformation across the Rio Grande rift near Socorro, New Mexico. J. Geophys. Res., v. 85, 7215-7220.

Sbar, M. L., and S. M. DuBois, 1984, Attenuation of intensity for the 1887 northern Sonora, Mexico earthquake. Bull. Seismol. Soc. Am., v. 74, 2613-2678.

Schlue, J. W., P. J. Singer, and C. L. Edwards, 1986, Shear wave structure of the upper crust of the Albuquerque-Belen basin from Rayleigh wave phase velocities. J. Geophys. Res., v. 91, 6277-6281.

Schmucker, U., 1964, Anomalies of geomagnetic variations in the southwestern United States. J. Geomagn. Geoelec., v.

15, 193.
Schmucker, U., 1970, Anomalies of geomagnetic variations in the southwestern United States. Bull. Scripps Inst. Oceanogr., v. 13, 33-53.
Scott, G. R., 1975, Cenozoic surfaces and deposits in the southern Rocky Mountains. In: Cenozoic History of the Southern Rocky Mountains, B. M. Curtis (ed.), Geol. Soc. Am. Memoir, 144, 227-248.
Seager, W. R., 1981, Geology of Organ Mountains and southern San Andres Mountains, New Mexico. New Mexico Bureau of Mines and Mineral Resources Memoir 36, 97 p.
Seager, W. R., R. E. Clemons, J. W. Hawley, and R. E. Kelley, 1982, Geology of northwest part of Las Cruces 1° x 2° sheet. New Mex. Bur. Mines Miner. Resour. Geologic Map 3, scale 1:125,000.
Seager, W. R., and G. H. Mack, 1986, Laramide paleotectonics in southern New Mexico. In: Paleotectonics and Sedimentation in the Rocky Mountain Region, J. A. Peterson (ed.), Am. Assoc. Petrol. Geol. Memoir, in press.
Seager, W. R., J. W. Hawley, F. E. Kottlowski, and S. A. Kelley, 1987, Geology of east half of Las Cruces and northeast El Paso 1° x 2° sheets. New Mexico Bur. Mines Miner. Resour. Geologic Map 57, scale 1:125,000.
Seager, W. R., and McCurry, M. O., 1988, The cogenetic Organ cauldron and batholith, south central new Mexico: evolution of a large-volume ash flow cauldron and its source magma chamber: Journal of Geophysical Research, v. 93, p. 4421-4433.
Seager, W. R., and P. Morgan, 1979, Rio Grande rift in southern New Mexico, west Texas, and northern Chihuahua. In: Rio Grande Rift: Tectonics and Magmatism, R. E. Riecker (ed.), Am. Geophys. Union, Spec. Pub., 87-106.
Seager, W. R., M. Shafiqullah, J. W. Hawley, and R. Marvin, 1984, New K-Ar dates from basalts and the evolution of the southern Rio Grande rift. Geol. Soc. Amer. Bull., v. 95, 87-99.
Shannon, J. R., Naeser, C. W., DeWitt, E., and Wallace, A. R., 1987, Preliminary summary of the timing of Cenozoic magmatism and tectonism in the Sawatch Uplift and northern Rio Grande Rift, Colorado (abs.). Geological Society of America Abstracts.
Sinclair, W. F., 1963, Geology of the upper East Fork of the San Juan River area, Mineral and Archuleta counties, Colorado. Colorado School of Mines, Unpublished M.S. Thesis.
Sinno, Y. A., P. H. Daggett, G. R. Keller, P. Morgan, and S. H. Harder, 1986, Crustal structure of the southern Rio Grande rift determined from seismic refraction profiling. J. Geophys. Res., v. 91, 6143-6156.
Sinno, Y. A., and G. R. Keller, 1986, A Rayleigh wave dispersion study between El Paso, Texas and Albuquerque, New Mexico. J. Geophys. Res., v. 91, 6168-6174.
Smith, D. L., and R. L. Jones, 1979, Thermal anomaly in northern Mexico: An extension of the Rio Grande rift? In: Rio Grande Rift: Tectonics and Magmatism, R. E. Riecker (ed.), Am. Geophys. Union, Spec. Pub., 269-278.
Steinpress, M. G., 1981, Neogene stratigraphy and structure of the Dixon area, Espanola basin, north-central New Mexico--summary. Geol. Soc. Amer. Bull., v. 92, 1023-1026.
Steven, T. A., 1975, Mid-Tertiary volcanic field in the southern Rocky Mountains. Geol. Soc. Am. Mem., v. 144, 75-94.
Stover, C. W., B. G. Reagor, and S. T. Algermissan, 1988, Seismicity map of New Mexico. U.S. Geol. Surv. Misc. Field Studies Map MF-2035, scale 1:1,000,000.
Swanberg, C. A., 1979, Chemistry of thermal and non-thermal groundwaters in the Rio Grande rift and adjacent tectonic provinces. In: Rio Grande Rift: Tectonics and Magmatism, R. E. Riecker (ed.), Am. Geophys. Union, Spec. Pub., 279-288.
Swift, C. M., 1967, A magnetotelluric investigation of an electrical conductivity anomaly in the southwestern U.S. Ph.D. Dissertation, MIT, Cambridge, MA.
Taylor, B. B., 1975, Neogene tectonism in south-central Colorado. In: Cenozoic History of the Southern Rocky Mountains, B. M. Curtis (ed.), Geol. Soc. Am. Memoir 144, 211-226.
Taylor, B., 1981, Heat flow studies and geothermal exploration in western Trans-Pecos Texas. Ph.D. Dissertation, Univ. of Texas at El Paso, 326 p.
Towle, J. N., 1980, New evidence for magmatic intrusion beneath the Rio Grande rift. New Mexico Geol. Soc. Am. Bull., v. 91, 626-630.
Tweto, Ogden, 1975, Laramide (Late Cretaceous-Early Tertiary) orogeny in the Southern Rocky Mountains, In: Geological Society of America Memoir 144, Curtis, B. F. (ed.), p. 1-44.

Tweto, O., 1978a, Tectonic map of the Rio Grande rift system in Colorado. In: Guidebook to Rio Grande Rift in New Mexico and Colorado, J. W. Hawley (ed.), New Mexico Bur. Mines Miner. Resour. Circ. 163, Sheet 1.

Tweto, Ogden, 1978b, Leadville mining district, Lake County, Colorado. In: Professional Contributions of the Colorado School of Mines, D. R. Shaw (ed.), 9, p. 68-71.

Tweto, Ogden, 1978c, Leadville district, Colorado. In: Ore Deposits of the United States, 1933-1967, J. D. Ridge (ed.), American Institute of Mining Metallurgical and Petroleum Engineers, Inc., p. 681-705.

Tweto, O., 1979, Geologic map of Colorado. U.S. Geol. Surv., scale 1:500,000.

Tweto, O., and Case, J. E., 1972, Gravity and magnetic features related to geology in the Leadville 30-minute quadrangle, Colorado. U.S. Geol. Surv. Prof. Paper 726-C, 31 p.

Upson, J. E., 1941, The Vallejo Formation: New early Tertiary red beds in southern Colorado. American Journal of Science, v. 239, p. 577-589.

Van Alstine, R. E., 1968, Tertiary trough between the Arkansas and San Luis, Colorado. U.S. Geological Survey Professional Paper 600-C, p. C158-C160.

Wallace, S. R., Muncaster, N. K., Jonson, D. C., MacKenzie, W. B., Bookstrom, H. H., and Surface, V. E., 1978, Multiple intrusion and mineralization at Climax, Colorado. In: Ore Deposits of the United States 1933-1967, J. D. Ridge (ed.), American Institute of Mining, Metallurgical and Petroleum Engineers, Inc., p. 605-640.

Warren, R. G., A. M. Kudo, and K. Keil, 1979, Geochemistry of lithic and single crystal inclusions in basalt and characterization of the upper mantle-lower crust in the Engle basin, Rio Grande rift, New Mexico. In: Rio Grande rift: Tectonics and Magmatism, R. E. Riecker (ed.), Amer. Geophys. Union Spec. Publication, 393-415.

Weidelt, P., 1975, Inversion of two-dimensional conductivity structures. Phys. Earth Plan. Interiors, v. 10, 282-291.

Wells, S. G., K. I. Kelson, and C. M. Menges, 1987, Quaternary evolution of fluvial systems in the northern Rio Grande rift, New Mexico and Colorado. In: Quaternary Tectonics, Landform Evolution, Soil Chronologies, and Glacial Deposits--Northern Rio Grande rift of New Mexico, C. M. Menges (ed.). Friends of the Pleistocene, Rocky Mountain Cell Guidebook, 55-59.

Wen, C. L., 1983, A study of bolson fill thickness in the southern Rio Grande rift, southern New Mexico, west Texas, and northern Chihuahua. M.S. Thesis, Univ. Texas at El Paso, El Paso, 74 p.

Williams, M. L., 1987, Stratigraphic, structural, and metamorphic relationships in Proterozoic rocks from northern New Mexico. New Mexico Bur. Mines Miner. Resour. Open-file Rept., 138 p.

Wolberg, D. L., Lozinsky, R. P., and Hunt, A. P., 1986, Late Cretaceous (Maastrichtian-Lancian) vertebrate paleontology of the McRae Formation, Elephant Butte area, Sierra County, New Mexico. New Mexico Geological Society, 37th Field Conference Guidebook, 227-234.

Woodward, L. A., J. F. Callender, and R. E. Zilinski, 1975, Tectonic map of the Rio Grande rift. New Mexico. Geol. Soc. Am., Map MC-11.

Woodward, L. A., J. F. Callender, W. R. Seager, C. E. Chapin, J. C. Gries, W. L. Shaffer, and R. E. Zilinski, 1978, Tectonic map of Rio Grande rift region in New Mexico, Chihuahua, and Texas. In: Guidebook to Rio Grande Rift in New Mexico and Colorado, J. W. Hawley (ed.), New Mexico Bur. Mines Miner. Resour. Circ. 163, Sheet 2.

Zietz, I., and J. R. Kirby, 1972a, Aeromagnetic map of Colorado. U.S. Geol. Surv. Geophys. Invest. Map GP-836, scale 1:500,000.

Zietz, I., and J. R. Kirby, 1972b, Aeromagnetic map of Colorado. U.S. Geol. Surv. Geophys. Invest. Map GP-880, scale 1:1,000,000.

Zoback, M. L., and M. D. Zoback, 1980, State of stress in the conterminous United States. J. Geophys. Res., v. 85, 6113-6156.

Zoback, M. L., R. E. Anderson, and G. A. Thompson, 1981, Cainozoic evolution of stress and style of tectonism of the Basin and Range Province of the western United States. Philos. Trans. R. Soc. London, Ser. A, 300, 407-434.

Zohdy, A. A. R., L. A. Hershye, P. A. Emergy, and W. D. Stanley, 1971, Resistivity sections, upper Arkansas River Basin, Colorado. U.S. Geol. Surv., Open-file Rpt., 71-002, 16 p.

